# Equipping Your Horse Farm

a live PTO (w/ shield or guard)

Category 2 hitch?

R1 or R4 tires → radials?

4 WD

# Equipping Your Horse Farm

## Tractors, Trailers, Trucks & More

Cherry Hill and Richard Klimesh

Storey Publishing

Edited by Lisa Hiley

Art direction by Vicky Vaughn

Cover design by Kent Lew

Text design by Kristy MacWilliams

Text production by Zuppa Design

Cover photograph by Richard Klimesh

Back cover photos by Cherry Hill &
Richard Klimesh (top); EquiSpirit
Trailers (bottom)

Author photo by Randy Dunn

Interior photographs by Cherry Hill and
Richard Klimesh, except for Ilona
Sherratt, page 46 middle; EquiSpirit
Trailers, page 79; Turnbow Trailers,
page 114

Illustrations by Terry Dovaston except
for pages 80, 112, and 113 by
Richard Klimesh

Indexed by Susan Olason

*The mission of Storey Publishing is to serve our customers by publishing practical information that encourages personal independence in harmony with the environment.*

The information in this book is true and complete to the best of our knowledge. All recommendations are made without guarantee on the part of the author or Storey Publishing. The author and publisher disclaim any liability in connection with the use of this information. For additional information please contact Storey Publishing, 210 MASS MoCA Way, North Adams, MA 01247.

Storey books are available for special premium and promotional uses and for customized editions. For further information, please call 1-800-793-9396.

Printed in the United States by Versa Press
10 9 8 7 6 5 4 3 2 1

**Library of Congress Cataloging-in-Publication Data**

Hill, Cherry, 1947–
    Equipping your horse farm / Cherry Hill and Richard Klimesh.
        p. cm.
    Includes index.
    ISBN-13: 978-1-58017-843-3; ISBN-10: 1-58017-843-X (pbk. : alk. paper)
    ISBN-13: 978-1-58017-844-0; ISBN-10: 1-58017-844-8 (hardcover : alk. paper)
    1. Horses—Transportation—Equipment and supplies. 2. Farm tractors.
    3. Horse Trailers. I. Title.
SF285.385.H54 2006
636.1'083—dc22
                                                                    2006008204

*To Ron Lonneman, friend and guru
of all things tractor and implement*

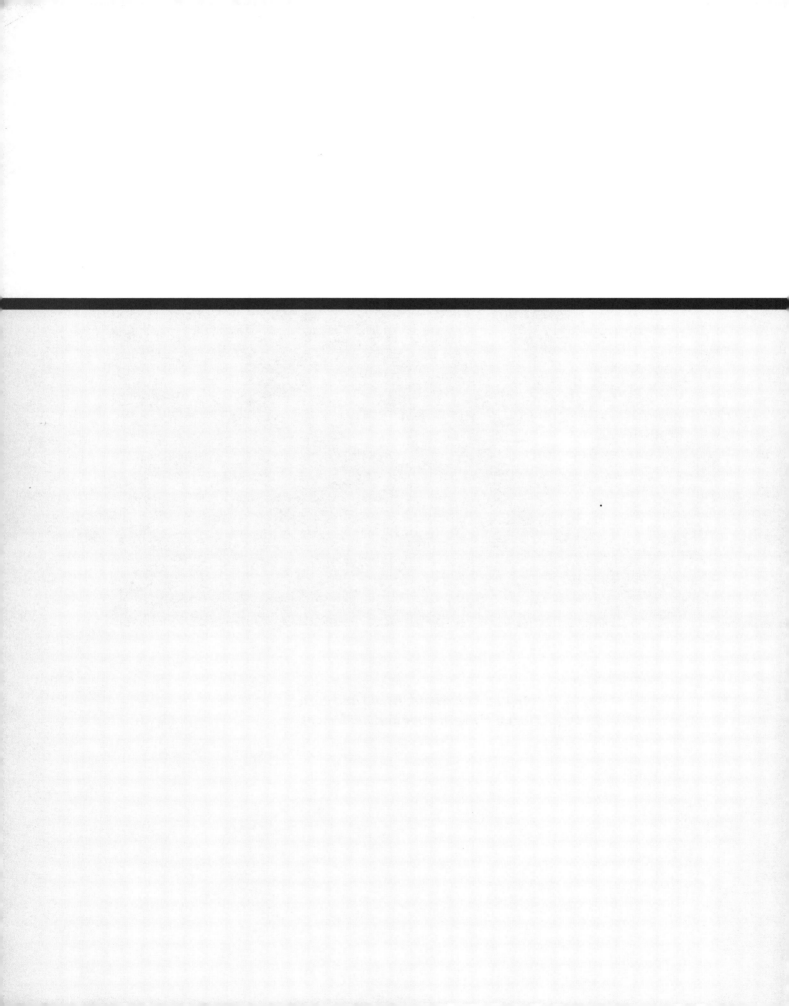

# Contents

# Preface

It's March, when we do our annual spring pasture grooming. We just spent six hours on our tractors and everything went like clockwork. Richard preps the pastures for me. He drives our garden tractor and a small wagon through all of the pastures, gathering rocks and consolidating them into out-of-the-way piles. He also moves manure piles away from fence lines so I can harrow the manure without taking out fence posts.

After Richard finishes, I come in with our utility tractor, pulling a spike-tooth harrow. The harrow distributes manure evenly throughout the pasture while smoothing out gopher mounds and other bumps and lumps. Then I attach a PTO-driven rotary mower to the tractor and trim any dried grasses that were left by the horses. The mower chops the grass into a fine mulch that provides nice protection from our pending heavy spring snows and rains.

It is a great feeling when everything goes so smoothly. Some farmers say that it is just good luck when machinery cooperates, but you can improve your luck by selecting the right tractor for the job, choosing appropriate implements, and taking good care of all your equipment. The same goes for the choice, use, and care of your truck and horse trailer. It is our hope that the information in this book will help you have many of those satisfying, productive days when everything just clicks along on your horse farm or ranch.

— Cherry Hill and Richard Klimesh

# Acknowledgments

Many thanks to:

Ron Lonneman of Ron's Equipment for his tireless help with photos and the manuscript

Kay Lonneman for letting me photograph *and ride* her ATV

J.T. Grainger for setting up countless tractors for us to photograph

Edgar Seaworth for allowing us to photograph his wide variety of equipment

Irv Fosaaen of Waukon Harley-Davidson for proofing the APV chapter

Stan and Roy Brown of Brown's Excavation for doing a great job over the years moving dirt and providing wonderful photo ops

Sam Shoultz, Ken Matzner, Curtis Conklin, and Molly A. Farrand of KESA Quarter Horses for helping us with many in-use photos

Richard R. Peters and Jerry Hubka for being models extraordinaire

# Your Needs

Although you will never eliminate the need for a basic wheelbarrow and a manure fork, those two essentials are only the beginning of the equipment needed to run a horse facility. Depending on the size of your operation and its associated tasks, several other kinds of equipment can help get the work done efficiently.

The equipment necessary to run even a small horse farm typically falls into one of two categories: a tractor or all-purpose vehicle with attachments and a truck with a trailer. You have likely already considered various options such as doing your farm chores by hand, buying your own machinery and hiring laborers to do the work, or hiring someone with equipment to do the work for you. If you don't have a trailer, you may also have considered hiring someone to haul your horses for you.

As you think about buying equipment for your place, and balance your budget with your wish lists, keep suitability and compatibility in mind. Consider joint ownership with a friend or neighbor, leasing equipment as you need it, or buying used items.

The initial cost of equipment (and of any necessary repairs to get it up and running, if it is a used item) is just the beginning of budgetary considerations. There will be regular maintenance costs for routine servicing and irregular repair costs (the *what* broke?). With some used equipment, repairs are more regular than irregular! There will be operating costs such as fuel and lubricants, and associated costs such as interest, insurance, and shelter. And, if you're like most tractor, truck, and trailer owners, there will be a never-ending list of "toys" you just have to have to use with your vehicles and equipment.

That's why it is so important to make wise choices on the key items — tractor and truck — that dictate the implements you can use with them. In this case, you can and should put the cart before the horse by listing all the chores you will need to do with your tractor and truck before you purchase those vehicles.

Keep suitability in mind as you shop. You want to buy the right implement the first time so you won't lose money "trading up." Stay within your budget, but try not to cut corners too much or you may end up with a vehicle that is difficult to resell. You should consider ease of operation, dependability, and warranty — they all add to peace of mind and, in the long run, are time- and cost-efficient. Here's where working with a knowledgeable, reputable dealer really pays off.

**Our neighbor's 6 x 6 utility task vehicle is perfect for his Colorado foothills acreage.**

# What Do You Need?

If the activities on your acreage justify investing in a truck, tractor, and attachments, take plenty of time to educate yourself on the choices and research the market so you can buy the equipment best suited to your needs.

Throughout this book, you'll find charts, lists, and worksheets to help you figure out what you need and checklists that you can take on buying trips. To start out, here is a general idea of how you can use a tractor around your farm or ranch:

- Collect manure
- Spread manure
- Harrow fields
- Mow weeds
- Clear brush
- Haul bedding
- Haul and move feed
- Maintain arena
- Plow snow
- Grade and maintain roadways
- Maintain pens
- Move portable shelters and large feeders
- Dig postholes
- Pull out old posts
- Power a water pump or generator
- Spray, fertilize, and seed pastures
- Cut, bale, and stack hay
- Cultivate soil for crops or gardens

In addition to a tractor, a truck and horse trailer is a handy rig for almost any horse owner and a must for anyone with a horse farm or ranch. You will need a truck to haul feed and bedding, carry fencing and building materials, and tow your trailer or other vehicles. Even if you don't take your horses off your property for shows or distant trail rides, you should have the means to transport them to a veterinary clinic in an emergency or to evacuate them in a flood or fire. Even for routine veterinary care, you can often avoid paying a farm call charge by taking your horse to the veterinarian's clinic. You may need to take a mare to a breeding facility or bring a new horse home. A trailer also comes in handy for hauling large quantities of hay and feed, fencing and building materials, and even furniture.

## WHAT KIND OF FACILITY?

Here are some typical scenarios that can help you get an idea of what you might need. Where do you fit?

---

*Jack and Jan Meek*
Two acres in California
One horse for their daughter

---

The Meeks are in their mid-thirties and their daughter, Sasha, is 14. Sasha has one horse that she takes to Pony Club rallies and trials. One acre is for the house, garage, yard, and the other acre has a small barn with all-weather pen, a sand dressage ring, and two small paddocks. The Meeks have manure picked up twice a week by a refuse company that has a contract with their homeowner's association; therefore, they don't require manure-spreading equipment. Here is what they use to maintain their acreage:

- Garden tractor with belly mower for yard and paddocks
- Wagon to pull behind tractor for moving feed, supplies, tack, and equipment
- Four-foot English harrow for grooming the pasture and arena
- Large SUV
- Two-horse, straight-load tagalong trailer with dressing room

**A garden tractor is handy for small jobs.**

### Bud and Bobbie Adamark
Ten acres in New York
Three senior horses

The Adamarks are retired. Their children, long since moved away from home, have left three senior horses on their parents' farm. The horses live on pasture nine months of the year (requiring feed during five to six of those months) and in large pens during the three months of late winter/early spring when the pastures are muddy. The Adamarks have one stacker load of alfalfa hay delivered each year and they purchase grass hay from a neighboring farmer in horse-trailer loads as needed. There are run-in sheds for shelter both in the pastures and in conjunction with the pens. Bud cleans out the sheds and pens with the loader on his tractor. Here is what they use to maintain their acreage:

- Subcompact tractor with loader, but no power take-off (PTO), three-point hitch, or rear hydraulics; used for clearing snow and taking feed to pasture horses during snowy weather
- Gas-powered mower for the pasture, towed by tractor
- Friction-drive spreader
- Riding lawn mower for mowing the yard
- Electric golf car with added bed for feeding on pasture during dry weather, moving light supplies, and taking guests out for a tour
- Large SUV
- Basic two-horse steel trailer for hay hauling and emergency horse transport

### Buck and Sissy Cole
Twenty acres in Minnesota
Four or five trail horses

The Coles are avid trail riders. They are in their late 20s and have no children. Their land is heavily wooded, but it has some good pasture that is usable for four to five months of the year. The pasture is covered in snow the rest of the year, so although the horses are often turned out for exercise in the winter, they also have stalls and covered pens. The Coles also have a large hay barn. Since their trail-riding season is short, Sissy has invested in an indoor arena so their horses can stay in shape year-round. Buck would like to clear some of the trees and brush for more pasture. To keep their acreage clicking, they have gathered up the following equipment:

- Compact tractor with loader, PTO, three-point hitch, and rear hydraulics
- PTO-driven spreader and mower
- Spike-tooth harrow for the pastures
- Rotary harrow for the arena footing
- Snowblower attachment for the tractor to keep lanes and areas around the buildings clear
- Three-point, PTO-driven wood chipper attachment for the tractor to make mulch of cleared brush and small trees
- Three-point, PTO-driven stump-grinder attachment to remove stumps of larger trees
- ¾-ton diesel truck
- Two-horse gooseneck trailer with living quarters
- 16-foot gooseneck flatbed trailer for hauling hay

**Our utility tractor is ideal for all of our ranch-related tasks.**

## Corky and Becky McCue
40-acre training facility in Texas
Twenty to twenty-five horses

The McCues are successful reining horse trainers who have just signed a five-year lease on a 40-acre facility with a 20-stall barn, indoor arena, outdoor reining track, hay barn, 10 covered pens, and one 32-acre pasture. Their lease agreement calls for normal maintenance plus subdividing the pastures and installing a permanent round pen and outdoor arena. They have two hired hands to help with the daily barn work and chores and another in charge of the fencing projects. They have hay delivered by the semi load. To run this facility, Corky and Becky have invested in the following equipment:

- Utility tractor with loader, PTO, three-point hitch, and rear hydraulics
- PTO spreader
- Six-foot, three-point mower with PTO
- Combination drag for conditioning arenas and reining track
- Eight-foot three-point blade for maintaining the long driveway
- Skid loader with bucket for cleaning pens and pallet fork for moving hay
- Auger attachment on skid loader for postholes
- Electric UTV for use at home and at shows
- Six-wheel gas UTV with dump bed for feeding on pasture, and hauling supplies and materials
- One-ton dually truck
- Gooseneck aluminum six-horse trailer with living quarters, tack room, and ramp to load UTV

**Hay-making equipment includes a mower.**

## Robert and Mary Broom
160-acre breeding farm in Iowa
Thirty to forty horses

The Brooms produce grass and alfalfa hay for their own use and for sale. Their herd consists of a stallion, 15–20 broodmares, the current year's foals, and about 10 young horses (yearlings and two-year-olds). They do not ride or train horses. They sell most of their young stock at a local production sale and the rest by private treaty at their farm. To run the farm, they have:

- Farm tractor with loader, PTO, three-point hitch, front and rear hydraulics, and cab
- Sickle bar mower for hay
- Rake for hay
- Baler for hay
- Disc, harrow, seeder for maintaining hay fields
- Three-point sprayer with booms for applying fertilizers, insecticides, herbicides
- Flex-wing rotary mower for pasture
- Large PTO spreader
- Front-mount hydraulic auger for fence/pen repairs and modifications
- Compact tractor with loader for cleaning pens and other areas
- All-terrain vehicle (ATV) for Mary to check on mares and foals
- Weed-sprayer attachment with hand wand for use on the ATV to spot-spray thistles and other weeds
- One-ton dually truck
- Stock trailer to transport horses to the sale
- Large SUV
- Two-horse, enclosed trailer with open floor plan that allows a mare and foal to be transported untied

Now you have an idea of the various ways farm vehicles and equipment can make horsekeeping easier. It's time for the fun part — shopping! To help you make good decisions, we've corralled all the information you'll need to consider before you purchase.

# 2 Tractor Features

As you search for your perfect tractor, you will make a number of "either/or" choices and ask many of those difficult "do I really need this?" questions. Whether you are looking for a new or used tractor, there are many things to consider.

Before you start, ask yourself whether your choice of tractor is driven by nostalgia or function. Although older tractors have a great sense of character and history, they often are tough to operate and maintain. Newer tractors have many excellent features that make them much more user-friendly and kinder to the land and environment. The U.S. Occupational Safety and Health Administration (OSHA) requires new tractors to have certain safety equipment and features that were not available on older tractors. If you are leaning toward owning and restoring a classic, realize that you might want to retrofit it with safety equipment, which can be costly and could destroy the classic look. For regular use on a horse farm or ranch, choose a reliable, functional tractor.

# Tractor Size

Tractors are generally grouped into four categories dictated by size, weight, horsepower, and suitability for purpose. If only manufacturers would agree on the size classifications. One tractor manufacturer calls its compact tractor a utility tractor while another calls its garden tractor a compact tractor. For the purposes of this discussion, I have divided tractors into four groups: lawn and garden, compact, utility, and farm.

### LAWN AND GARDEN TRACTORS

These small tractors generally have either a one- or two-cylinder gas engine or a two- or three-cylinder diesel engine that provides up to 25 hp. They weigh between 500 and 1200 pounds.

A lawn tractor is designed for mowing lawns. It has tires that minimize footprints (tracks) on lawns and a two- to three-gallon fuel tank. It might have an electric power take-off (see page 11), but typically can accommodate few attachments, if any. A garden tractor has tires with good traction for working in a garden, a five- to six-gallon fuel tank, and a PTO (electric powered or engine powered) with a hydraulic clutch. Some offer four-wheel drive. Depending on the brand, garden tractors can be used with a full line of compatible attachments such as a dump cart, trailer, tiller, broadcast spreader, snow thrower, blade, roller, sprayer, spike aerator, disk, plow, cultivator, and rotary broom.

## What About an APV?

You might discover that you want to purchase an all-purpose vehicle instead of, or in addition to, a tractor. Read chapter 5 to learn how APVs differ from tractors. Many of the features and implements are similar, so even if you are not considering an APV, you'll find lots of useful information there.

A lawn or garden tractor can be useful for cleaning stalls or pens, or for feeding a few horses at a time, because it can maneuver in tight spaces while pulling a small manure or feed cart. It is unsuitable, however, for large-scale feeding or manure handling, routine fieldwork, or maintaining an arena. Smaller tractors can be an expensive option for horsemen. By the time you buy one and add a cart and other attachments, you probably could have purchased a compact tractor. You don't want to end up with half the tractor power for the same price. Depending on the scale of your operation, a lawn tractor for light duty or barn work might make sense only if you are planning to buy two tractors.

### COMPACT TRACTORS

This category is often broken into subcompact and compact tractors. These convenient and easy-to-operate tractors are also called mid-size or acreage tractors. They provide 25–45 hp and can have a two-

**Lawn tractor**

**Garden tractor**

or a four-cylinder engine and a Category 1 or 2 hitch. They weigh between 1400 and 2000 pounds. Since they are not very tall, they are pretty easy to mount. The hitch is low to the ground, making attachment of implements convenient, yet with compacts there is greater ground clearance (12 inches or more) than with garden tractors or subcompacts. Compact and subcompact tractors are typically good tractors for novice drivers.

Subcompacts are available as low as 15 hp and up to 25 hp (10–16 PTO hp) with two- or three-cylinder diesel engines. They differ from garden tractors in features, style, and fuel. These models are only available with diesel engines. Some subcompacts have four-wheel drive, a Category 1 three-point hitch, a loader, and a wider range of beefier implement choices. Their fuel-tank capacities are six to eight gallons.

A compact tractor usually has a three- or four-cylinder diesel engine with up to 45 hp and uses Category 1 or 2 equipment (see Hitches on page 14). These tractors work well for all-around chores on smaller acreage, such as cleaning out pens and runs with a front-end loader or a six-foot blade on the back, but they are limited by the size of the attachments they can handle. They can pull an eight-foot pull-type disc or a six-foot three-point disc, which is suitable for a large arena or a small field. They also work well with a small manure spreader, especially the friction-drive type.

## UTILITY TRACTORS

These taller, more powerful tractors offer approximately 45 to 80 hp with a three- to five-cylinder engine and a 20–30-gallon fuel tank. Weighing 1500–3000 pounds and taking a Category 2 hitch, they are able to operate heavy-duty equipment, such as a posthole digger or a large loader to scoop or push deep snow. If you have a big arena and want to use a disc with two eight-foot sections, you will want to consider a utility tractor. If you regularly handle large amounts of manure with a heavy-duty, PTO-driven spreader, you will need a more powerful tractor like this. If you shop around, you may be able to find a utility tractor for the same price that you would pay for a compact tractor. All other things being equal, buy the larger tractor if you think you will ever need the extra power.

## FARM TRACTORS

Because these tractors are designed for commercial farming, they are very large and powerful, providing more than 85 hp (as much as 450 hp) with a four- to six-cylinder engine and a fuel tank with a capacity as large as 300 gallons. They weigh between 2500 and 6000 pounds, can take a Category 2, 3, 3N, 4, or 4N hitch, and have many features. They often have multiple hydraulic hookups and PTOs at the rear, front, and side. The transmission could have as many as 24 forward and 24 reverse gears with on-the-go four-wheel drive.

**Subcompact tractor**

**Compact tractor**

The tires are so big that the trip to the cab may take two or three steps, but the climb is well worth it. Large tractor cabs frequently have all the comforts of home: a plush power seat with multiple adjustments, a heating and air conditioning system, sound system, tilt steering wheel, GPS system, and more. Cabs on new tractors will have a certified rollover protective structure (ROPS), but this is not necessarily so on older models. New cabs are well sealed to keep out dirt, dust, and fumes. Farm tractors without cabs might have a foldable ROPS that allows you to park under a lower roof.

This is an EROPS (Enclosed Rollover Protective Structure) certified cab.

# Choosing an Engine

This decision is pretty easy because since 1974, all tractors (except for some lawn and garden tractors that come with one- to three-cylinder gas engines) manufactured in the United States have three- to six-cylinder diesel engines. Diesel engines are preferred because they are two to three times more economical to run and they produce more torque per cubic inch than a gasoline engine. Diesels are somewhat easier on the environment, because they emit less carbon monoxide and unburned hydrocarbons than gas engines and therefore produce lower greenhouse gas emissions. Diesels, however, have higher emissions of some pollutants, including nitrogen oxides and diesel particulates. New particulate filters and low-sulfur diesel fuel may assist diesel engines in

meeting the U.S. Environmental Protection Agency's standards for particulate matter and hydrocarbons. Your tractor dealer can update you on these options. (Read more about gas engines in chapters 5 and 7.)

## DIESEL ENGINES VS. GAS ENGINES

With more torque, diesel engines have higher load-carrying capacity and greater pulling power. In addition, a diesel uses 25 percent less fuel than a gasoline engine doing the same work. However, a diesel engine is heavier than a gas engine with comparable horsepower.

Diesels are more dependable and break down less often than a gas engine, but repairs generally cost more. Maintenance requirements, such as oil

**Utility tractor**

**Farm tractor**

changes, are higher for diesels, but since they use a fuel injection system instead of a carburetor, there is no need for tune-ups and spark plugs. And a diesel's exhaust system will last longer because diesel exhaust is less corrosive than gasoline exhaust.

Choosing a diesel engine requires that you store diesel fuel on your property. If you already have gas on hand, say for a lawn mower or all-terrain vehicle, it is important to know that the two fuels are not interchangeable, so you will need to mark your containers to keep them straight. Diesel fuel is not as volatile as gasoline — it will burn but it will not explode. Diesel fuel gels in cold weather, so you'll need to buy a winter blend or put an additive in the fuel to prevent problems. Partly because of the fuel and partly because they use heavier oil, diesel engines can be harder to start in cold weather than gas engines. If you live in an area where the temperature drops below 10°F, an engine heater is a good idea to ensure cold weather starts.

Diesel engines have a much longer lifespan than gas engines, and diesel vehicles typically hold their value better than gas-powered ones. By the same token, diesel vehicles cost approximately 25 percent more than similar gas-powered vehicles. If you need an engine that will stand up to many years of hard use, the extra cost of a diesel engine is usually justified.

## Rollover Protective Structure (ROPS)

**According to the National Safety Council,** more than half of farm fatalities involve rollovers. That's why a rollover protective structure and seat belts are mandatory on all tractors built since 1984. A ROPS is like a roll bar in a jeep or other off-road vehicle. It is a framework that is strong enough to hold its shape even if the tractor is upside down with all its weight on the ROPS. The ROPS creates a safe cage or protected zone for the operator, providing the operator is wearing a seat belt.

Most Japanese tractors have safety features comparable to U.S. tractors but some other foreign-made tractors do not. Many used farm tractors, because of their age, do not have safety gear. If you purchase an older tractor or a foreign tractor, it is up to you to update your machinery and to maintain it and operate it in a responsible manner. All major equipment manufacturers provide tractor safety programs, including information about retrofitting older tractors with ROPS.

Not all cabs provide ROPS protection. Many cabs have a light frame and were designed to protect the operator from dust, noise, and the weather, rather than to prevent injuries. If a cab is ROPS-certified, it will have a plate or stamped designation on it.

Some tractors have a sunshade or canopy that carries the FOPS designation for falling object protective structure, which means it is sturdy enough to protect you from falling objects. Other sunshades just attach like an umbrella to the ROPS. In either case, a sunshade or canopy in itself does not offer any protection if the tractor rolls over. Only with a ROPS do you have adequate protection.

**Engine access by hood**

**Engine access by side panels**

### Engine Access

For routine servicing, you need easy access to the engine, filters, radiator, battery, and other components. Some tractors have hoods that raise up in front like a truck's, while others require the removal of side panels. Some side panels can be tricky, or even downright difficult, to remove and replace. Try them before you buy.

## SIZE AND HORSEPOWER

Choose a tractor for the work you will be doing and buy one with enough horsepower to run the implements you require. It is better to have more horsepower and not need it than to need more horsepower and not have it. Make sure the tractor will fit into the areas where you work, such as in a loafing shed or under a barn overhang, and into the storage shed where you will park it.

Engine size is written as either cubic inches (cu. in.) for larger engines in cars and trucks, or in cubic centimeters (cc) for smaller engines like those in ATVs and utility task vehicles (UTVs). Engine size is measured by the total displacement in all the cylinders of the engine. The displacement is determined by the bore (the diameter of the cylinder and the piston) and by the stroke (how far the piston travels up and down). The volume of the cylinder when the piston is at the very bottom of its stroke is the displacement of that cylinder. Generally, as the number of cylinders increases, power increases.

An engine can be naturally aspirated or turbocharged. In a naturally aspirated engine, the air or air-fuel mixture is drawn into the cylinders with atmospheric pressure. A turbocharged engine has a device like a powerful fan that forces greater amounts of air and fuel into the cylinders, thereby increasing power. A turbocharged engine delivers more power and torque at low engine speeds. Turbo charging is particularly advantageous for diesel engines and especially those operating at high altitudes.

# Power Take-off

A power take-off (PTO) is a revolving splined (interlocking) shaft that is an extension of the drive train and usually found on the back of a tractor; sometimes it is also on the front and/or side. It transfers power to machinery such as a manure spreader, auger, or mower by means of a removable drive shaft with a splined coupling. Optimum speed of a

## What Is Horsepower?

Horsepower is defined as the amount of energy or work required to raise a weight of 550 pounds a height of one foot in one second, or to overcome or create a force that is equivalent to doing that amount of work.

There are various ways to represent horsepower in machinery. Some tractor manufacturers use the gross horsepower rating of the engine, but this doesn't take into account the energy that is lost as it moves through the tractor's drive train to the wheels or PTO in order to be used. Others manufacturers use laboratory tests to take horsepower measurements, but because the tests are not "real world," their accuracy can be hard to assess.

In one test, a dynamometer is connected to the tractor PTO, and the engine is operated at various speeds against standard resistances. The values are averaged to give the PTO horsepower, or how much power the tractor can continuously deliver to an auxiliary piece of equipment through the PTO. The PTO horsepower is 75–98 percent of the gross engine horsepower.

Perhaps the most useful measure is sustained or continuous horsepower (how much work the vehicle can perform over a sustained period of time), measured by the dynamometer. To test a tractor's pulling ability, its wheels rest on the dynamometer platform while it is operated at varying speeds in all gear ranges against standardized resistances that simulate loads attached to the drawbar. The resulting averaged pulling capacity is known as the drawbar horsepower, or how much power the tractor can continuously deliver to the drive wheels. Drawbar horsepower is approximately 50–85 percent of the PTO horsepower.

PTO is 540 revolutions per minute on most utility and smaller tractors. On larger tractors, PTO speed might be stated as 540 or 1000 rpm.

A live or continuous PTO rotates independently of the tractor drive train. A live PTO continues running when you change speed and direction and allows you to power an implement with the engine declutched and the tractor standing still. With a semi-live or two-stage PTO, the shaft is engaged and disengaged in conjunction with the tractor drive train by operation of a two-stage clutch; therefore the tractor must be in motion for the PTO to operate. This is not as handy; a live PTO is much more desirable.

PTOs are extremely dangerous and great care should be taken when working around them. If clothing or hair get caught in the revolving shaft, serious injury or death is likely. Tractors and other equipment produced since the 1970s have mandatory shields or guards around dangerous parts of farm machinery like the PTO, but older models do not have as many safety features. If you use a PTO, even a shielded one, keep children away from the

equipment, and be very cautious when working with it yourself. If you have an unshielded PTO, the manufacturer of your tractor should have information on retrofitting the PTO with a shield, or you could have a blacksmith or welder fabricate guards for your machinery.

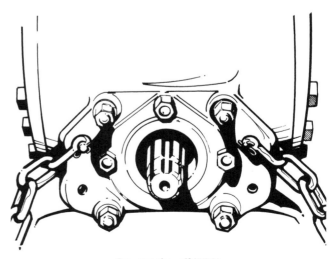

**Power take-off (PTO)**

# Transmission

A transmission is a device that uses gearing and torque conversion to change the ratio between engine rpm and driving wheel rpm. The transmission can make a tractor user-friendly or it can make the user tractor-angry. A tractor's transfer case (usually located to the right of the operator) is made up of at least two transmission shift levers. On many compact and utility tractors, the range shift lever provides four major gear ranges such as Low Low, Low, Medium, and High range. The gearshift lever provides four gear changes (1–4) with each range. This results in 16 forward and 16 reverse gear speeds. Some compacts only have eight forward and eight reverse gears. Lawn and garden and subcompact tractors have even fewer gears; farm tractors have more, as many as 24 forward and 24 reverse.

Manual or standard shift transmissions are durable, and ideal for pulling heavy loads, but they are not handy because you usually need to clutch and bring the tractor to a stop when shifting gears and ranges. If you shift while the tractor is moving, the two gears involved will grind because they are trying to mesh while rotating at different speeds.

A synchro-shift transmission is a type of standard transmission with synchronizers that minimizes or eliminates grinding when you shift between gears while the tractor is in motion. The clutch pedal must be depressed when you shift, but the tractor does not have to be stopped. Some synchro-shift transmissions are synchronized at all speeds, while others have limited synchronization.

A shuttle-shift transmission synchronizes forward and reverse and allows you to shift back and forth between the two with one movement while staying in the same transmission speed. You have to depress the clutch and operate a manual lever, but you do not have to bring the tractor to a stop. Shuttle shift is handy for repetitive back-and-forth operations like picking up a bucket of manure, backing up, and then moving forward to load it into a spreader.

A power shift transmission allows you to shift between different speeds and direction without

**Transmission range shift lever**

**Transmission gearshift lever**

**Shuttle shift lever**

using the clutch pedal — much like an automatic transmission in a truck. The shifting occurs from pressure on a foot pedal like an accelerator.

Hydrostatic transmissions are also controlled by a foot pedal and allow you shiftless acceleration. When you take your foot off the pedal, the tractor stops, which is a good safety feature. Often these have two pedals — the left one for forward motion and the right one for reverse.

*Note:* While most shifting takes places with floor pedals, the accelerators on tractors can be either in the form of a hand throttle on the column or an accelerator pedal on the floor or both.

## DIFFERENTIAL

A differential is a set of gears that transfers power from the transmission to the wheels. It also can slow the rotational speed from the transmission. A differential is designed to allow the wheels to rotate at different speeds, which enables a vehicle to turn without the wheels binding when the outside wheels need to turn faster than the inside wheels. One problem is that the differential also allows the wheel with the least traction to spin, as you might have experienced on snow or ice. On rough terrain, if one drive wheel comes off the ground, it will spin helplessly while the other wheel just sits there, and you're stuck.

A limited slip differential, sometimes called positraction, allows normal differential action when going around turns, but when a wheel spins, the differential transfers more power to the non-spinning wheel. This results in much better overall traction.

A locking differential can act like a solid axle to make both wheels turn at the same speed while receiving maximum power. If one wheel comes off the ground, the other one carries on — the left wheel doesn't care what the right wheel is doing.

This option can be standard or optional. On some tractors it is engaged with the wheels moving; on others the wheels must be stopped. Differential lock is operated manually by a lever or pedal, or electronically by a button or switch. Automatic differential lock engages when needed and unlocks for turning.

# Hitches

How and where you attach implements to your tractor will dictate what kind of work you can do and how efficient the tractor will be. Most tractors have a straight-bar hitch or pin hitch, also called a drawbar, which is a frame-mounted bar of steel at the rear of the tractor where implements can be attached. The drawbar is used for pulling trailers, wagons, and some discs and harrows. A hitch pin connects most implements, but a ball can be attached to the drawbar to pull implements or trailers with a coupler. (Read more about ball and coupler hitches in chapter 8.)

Many tractors also have a three-point hitch,

## Three-point hitch classifications

### CATEGORY 0

Tractors with up to 20 hp (most garden tractors)

Top-link pins are ⅝ inch in diameter

Lift-arm pins are ⅝ inch in diameter

Width of arm spread 19 inches when implement is centered

### CATEGORY 1

Tractors with 21–45 hp (most compact tractors)

Top-link pins are ¾ inch in diameter

Lift-arm pins are ⅞ inch in diameter

Width of arm spread 26 inches when implement is centered

### CATEGORY 2

Tractors with 46–94 horsepower (most utility tractors)

Top-link pins are 1 inch in diameter

Lift-arm pins are 1⅛ inch in diameter

Width of arm spread 32 inches when implement is centered

### CATEGORY 3

Tractors with 95 hp and more (most farm tractors)

Top-link pins are 1¼ inch in diameter

Lift-arm pins are 1⁷⁄₁₆ inch in diameter

Width of arm spread 31–33 inches when implement is centered

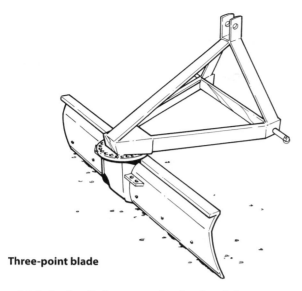

**Three-point blade**

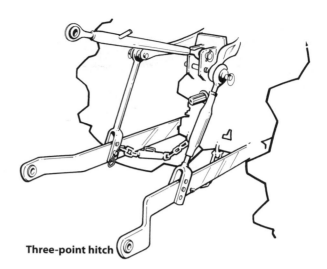

**Three-point hitch**

which is the linkage on the back of the tractor that uses hydraulic power to raise and lower equipment such as discs, plows, and posthole diggers. The two side arms, also called lift arms or draft links, serve to both lift and pull. The top arm, called the top link or center link, acts as a pivot point to raise the implement and dictates the angle of the implement to the ground. The added weight of the raised implement on the tractor's rear wheels provides increased traction.

Hitch categories refer to the size of the connecting pins and the strength of the components; the smaller the number, the smaller the hitch. Any Category 1 implement is compatible with a Category 1 hitch. Most modern compact tractors have either a Category 1 or 2 three-point hitch. Some Category 1 implements can be adapted to Category 2 hitches, but Category 2 implements cannot fit a Category 1 hitch. Utility tractors usually have a Category 2 hitch. Farm tractors usually have a Category 3 hitch or larger.

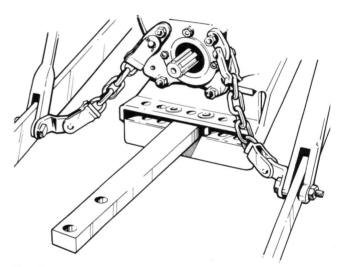

**Drawbar**

## HYDRAULICS

A hydraulic system is comprised of a reservoir of hydraulic fluid, hydraulic pumps, and hoses to deliver pressurized fluid to the connectors where implements are attached. A hydraulic system is rated by flow in gallons per minute and lift capacity at the hitch point or 24 inches behind it. Larger tractors have several separate pump systems (one for the

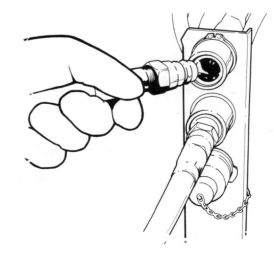

**Hydraulic hose connectors**

## HYDRAULIC SYSTEMS

| TYPE OF TRACTOR | GALLONS PER MINUTE | LIFT CAPACITY |
|---|---|---|
| Garden | 5 gpm | 500–1000 pounds |
| Compact | 4–11 gpm | 1200–2900 pounds |
| Utility | 12–20 gpm | 3200–8000 pounds |
| Farm | Over 20 gpm | Over 8000 pounds |

three-point hitch at the rear, one for steering, one for the loader, etc.) that maintain optimum operating and lifting power when running several implements at the same time.

# Other Features

Don't be overly influenced by a shiny appearance, as fresh paint can often cover up or distract from significant problems. Keep in mind that 90 percent of all tractors are stored outdoors. A tractor can look weather-beaten, with oxidized paint and ragged seat covers, yet be functionally sound. But there are other features you should consider, such as the condition of the tires and whether you need two- or four-wheel drive.

### TIRES

On most used tractors, the tires are quite old and you can expect to see some minor weather checking (cracking) from the deteriorating effects of sun,

**R1 Agricultural tread**      **R3 Turf tread**

mud, and water. If the tires are showing deep fissures, however, beware — they could come apart at any time. Generally speaking, tread depth is not as important on tractor tires as it is on your car or truck. Instead, look for lugs with at least 50 percent of their tread depth left. New 28-inch tires have one-inch tread depth and new 38-inch tires have 1½-inch tread depth.

Treads come in various types. One that is well suited for tractors used on horse acreages is the R1 agricultural tread, which has diagonal bars running from the center to the inside and outside edges of the tire. R3 (turf tires) and LSW (long, straight, wet: these are fine-turn, high-floatation tires) are wider and smoother, making them ideal for smooth, level pastures as they cause little damage to plants. R4 industrial tires are also popular; they support weight without digging in as deeply as the R1 tread. R4s are more like backhoe tires or a cross between a turf tire and an agricultural tire. They have bars that are wider and not as deep (aggressive) as those on agricultural tires.

### Bias Ply and Radials

A bias-ply tire has plies that lay in a criss-cross fashion across the centerline of the tread and has stiff sidewalls. This gives the tire strength, but it becomes hot and wears out more quickly than a radial tire. In a radial tire, the plies run directly across the tread and are combined with a rigid belt made of fiberglass or steel mesh that is located between the plies and the tread. This construction provides longer tread wear but a rougher ride. The belt overwrap of a radial tire causes less tread distortion under load and more sidewall distortion.

Radial tires for tractors can result in more efficient fuel use and increased towing efficiency, as well as reduced tire slip and less vibration. They may extend the tire life by as much as 2000 hours. On pickups, radial tires can produce increased sway, especially at highway speeds.

New tire size codes look something like this: 320/90 R 50, where 320 is the tire width in millimeters and 90 is the ratio of the height of the sidewalls to the tire width. In this example, the tire sidewalls are 90 percent as tall as the tires are wide. R equals the type of tire, which in this case is a radial, and 50 gives the rim code or size in inches; that is, the inside diameter of the tire and the outside diameter of the wheel where the tire bead sits.

Front tractor tires are in the same price range as truck tires. Two-wheel-drive front tires are about half the price of four-wheel-drive tires. Rear tractor tires, however, will cost substantially more. Although used rear tires will be half the price of new, the condition of the tires will greatly determine whether you are getting a good deal or not.

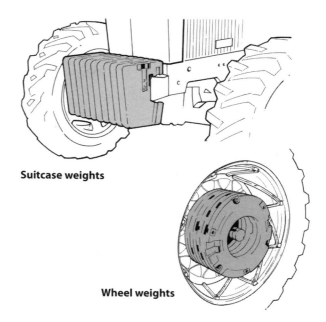

**Suitcase weights**

**Wheel weights**

## Tire Chains

Your regular tractor tires should provide enough traction under almost any conditions, but if you plow snow or operate in deep snow or icy conditions, especially in hilly terrain, tire chains will greatly increase your traction. They also tear up the land.

Tire chains come in many sizes to fit various tires. When spread out flat, tire chains look like a ladder made of chain. To apply the chains, lay them out in front of the wheels, drive the vehicle onto the chains, and then wrap the chains around the tire and connect the fasteners.

It's best to take tire chains off when they are not required since chains damage the ground more than bare tires do and wear out more quickly when used on dry or hard ground.

## ADDING WEIGHT

Tractors benefit from added weight in several situations. One example is when you are pulling a heavy implement, such as a disc, through deep ground, and the rear tires lose traction and start to spin. Adding weight to the rear can give more traction. When moving heavy material, such as gravel, with a loader, the tractor might become front heavy and lose traction at the rear, in which case it would also benefit from more rear weight.

With good traction at the rear, the front of the

tractor can actually lift up off the ground when pulling a heavy load, resulting in bouncing or a dangerous rear flip-over. Adding counterweights to the frame at the front of the tractor or filling the front tires with water or other fluids can help maintain balance and increase front traction.

Some tractors have weight brackets on the front, sides, or rear that allow you to add weight plates much as you would when stacking a barbell. You can also add metal weights to the rims of the rear wheels. Or you can fill the rear tires with water or other fluids. More weight at the rear increases the gross weight of the tractor, which increases the amount of horsepower delivered to the ground. This gives you more traction, but also increases the depth of the track of the rear tires, especially in soft ground. When working in wet or boggy terrain, a tractor with filled tires could get stuck more easily.

If you use water to fill tractor tires, be aware that it can freeze in the winter. You can add antifreeze, but it is expensive. Another option is calcium chloride solution, which is much cheaper than antifreeze. The tires are filled to three quarters of their volume with the solution and then inflated with air. The disadvantage of calcium chloride is that it is corrosive. However, if tires and wheels are well maintained, and leaks are spotted early and repaired, the rust can be kept at a minimum. Another option is one of the new non-salt products available. Your tractor dealer can order them for you.

## TWO-WHEEL DRIVE OR FOUR-WHEEL DRIVE

Although a substantially priced option, four-wheel-drive tractors are often the best choice because they are more suitable for rough or hilly work and have better traction in ice, snow, and mud. They also tend to hold their resale value better than two-wheel-drive models, which helps offset the initial investment. The heavier front-axle mechanism of a four-wheel-drive vehicle helps to even out weight distribution, which improves balance and stability, boosts pulling power, and increases traction for the use of a front-end loader. Four-wheel drive can have manual engagement (with a shifter), button engagement, or automatic engagement. If you do opt for a two-wheel-drive model, look for one that has a differential lock that locks the rear wheels together (like positraction), which will help keep you from getting stuck.

## POWER STEERING

Manual steering works fine for long straight lines and with smaller tractors, but if you are maneuvering in tight spaces, as you must when cleaning pens, power steering is essential. Overall handling is much improved with power steering on any tractor more than 35 hp.

## AGE AND MAKE

Wear on tractors is measured by hours of engine operation rather than by miles or by calendar age, so don't be overly concerned with the year a tractor was manufactured. If buying a used tractor, try to purchase one with less than 5000 hours on it. Many people have become accustomed to buying a new car or truck every few years, but that is false economy with tractors. The quality of workmanship is really fine on many older models, so if a tractor runs well and is paid for, any extra age just adds to its character!

Good equipment retains its value and, in some cases, appreciates. In 1948, a brand-new 8N Ford cost $900 right off the showroom floor. That same tractor is worth $2000 today. There are a lot of very usable, good tractors from the 1950s and '60s. Just be sure that parts are available for repair, and that you think carefully about upgrading any safety features that may be missing.

There are some types of tractors to avoid as practical work vehicles because they are so rare that it is difficult, if not impossible, to find parts for them. Any make of narrow front-end tractor falls into this category. Since these tricycle tractors were discontinued in the 1960s, you probably won't find too many of them in running condition anyway.

Because of a renewed interest in restoring old tractors, however, there are after-market companies offering parts and manuals for older tractors. These are not original equipment manufacturer parts, and because the market is for collectors and restorers, the price of parts is sometimes high, but they are more readily available than they have been previously.

Tractors without a three-point hitch, sometimes called dry-land tractors, are very impractical. Buying an old Minneapolis-Moline for $200 may seem like a great deal, but not after you realize that the draw bar on the back limits you to using a pull-type disc and that parts are very difficult to locate. Parts are generally available for most older models of New Holland, Massey Ferguson, John Deere, International Harvester, and Allis Chalmers, so these models would probably be a better choice.

## WARRANTY

New tractors are typically sold with a full two-year or 2000-hour warranty for parts and labor. Used tractors that are reconditioned and sold through a dealer might have a 90-day parts and labor warranty, but generally with used tractors it's a case of "buyer beware." It may be worth it to pay more for a reconditioned tractor in order to have the peace of mind and financial assurance that if anything goes wrong, it will be fixed. Most glitches seem to turn up within the first few weeks of operation.

# All About Batteries

A battery stores electricity for starting gas and diesel engines in vehicles including tractors and trucks. A group of batteries powers the engine in electric golf cars, ATVs, and UTVs.

The battery can also power other parts of the vehicle as well as certain attachments and tools — radio, dump box, winch, drill — without the engine running. All vehicles use at least one battery, so you should learn a bit about them in order to make informed purchases and to perform proper maintenance.

Vehicles use lead-acid batteries. This kind of battery has two types of lead plates surrounded by an electrolyte solution of dilute sulfuric acid (battery acid) to convert chemical energy into electrical energy and back again. Electrolytes are nonmetallic substances like acids and salts that conduct electricity when dissolved in water. The battery makes a negative charge at one post, or terminal, and a positive charge at the other post. When the posts are connected by a conductive material, electricity flows between them.

The battery case is made of high-density polypropylene or hard rubber and contains a number of individual cells; each holds lead plates and acid. Each cell is two volts and has an external cap, so you can tell the size of a battery by counting the caps and multiplying by two: six caps means a 12-volt battery.

The type of battery used in trucks and tractors is the flooded-cell battery. It has a liquid electrolyte covering the lead plates, so it must be kept upright to prevent the acid from spilling or leaking out.

The electrolyte in a battery evaporates and is used up over time. Many flooded-cell batteries have removable caps that allow you to add distilled water and electrolyte as needed.

**12-volt starter battery**

The caps on the cells of some flooded-cell batteries, however, are sealed, and those batteries are called "maintenance free." When the electrolyte in a sealed battery is used up, the battery is finished.

## TYPES OF BATTERIES

There are three types of lead-acid batteries; they look very similar but have different purposes.

Starter batteries are designed to provide a lot of power quickly in order to get a gas or diesel engine going, and then to recharge quickly. They do not tolerate frequent deep discharge (being drained beyond 80 percent of capacity). Most battery manufacturers recommend not discharging batteries more than 50 percent before recharging them. A starter battery can usually only withstand between 15 and 30 deep cycles before it needs replacing.

Deep-cycle, or deep-discharge, batteries are for powering electric engines such as those in golf carts. They have less instant energy but produce more energy over a long haul. As their name implies, they are designed to tolerate deep discharges, at least several hundred cycles, before they need replacing. They are generally heavier than starter batteries because they utilize thicker lead plates. You can use a deep-cycle battery for a starter battery; it just won't have as much initial oomph. You should never use starter batteries for deep-discharge applications like running electric engines.

Be aware that some marine/RV batteries are sold as dual-purpose batteries for both starter and deep-cycle applications. If used for deep-cycle tasks, they will only last about half as long as an actual deep-cycle battery.

# Tractor Spec Sheet

Now that you know a little more about tractors, it is time to decide on your perfect tractor. To help, you can use the accompanying flow chart to walk step-by-step through the choices. Then fill out your wish list on the next page.

**1** **Do you already have implements that you want to use with the tractor?**
- If yes, go to 2.
- If no, go to 3.

**2** **If you already have implements, the following will give you an idea of the size of tractor you will need to operate the implements.**
- For implements with a Category 0 three-point hitch, consider a lawn and garden tractor up to 20 horsepower.
- For implements with a Category 1 three-point hitch, consider a compact tractor up to 45 horsepower.
- For implements with a Category 2 three-point hitch, consider a utility tractor up to 85 horsepower.
- For implements with a Category 3 three-point hitch or larger, consider a farm tractor with more than 85 horsepower. Go to 4.

**3** **If you are buying both a tractor and implements at the same time, plan for your future needs.**
- If your needs are limited, your facilities are compact, and your property is less than two acres, consider a garden tractor or subcompact.
- If your needs are moderate, and your property and facilities are medium-sized, up to 20 acres or so, consider a compact tractor.
- If your needs include some pasture work and loader and spreader work, and your facilities are large or spread out over 20 acres or more, consider a utility tractor.
- If you grow grain crops, make hay, or have a large operation, consider a farm tractor.

**4** **What will you use your tractor for?**
- To perform daily feeding and cleaning in small spaces, or to mow in tight spaces consider a lawn or garden tractor.
- To mow small pastures and do basic small acreage work, consider a compact tractor.
- To operate a PTO-driven manure spreader or posthole digger, or to perform heavy loader work, consider a utility tractor.
- To farm crops, make hay, and feed large groups of horses, choose a farm tractor.

**5** **Do you plan to keep your tractor for more than five years?**
- If so, consider a new diesel tractor.
- If not, consider a used diesel tractor or a used gas tractor.

**6** **Are you color blind? Do you just have to have that red or green or blue tractor?**
- If so, list the brand on your worksheet.
- If not, shop around. Your open mind may lead you to a better bargain.

**7** **Do you require a certain type of transmission? If so, list it here:**

_____

_____

**8** **Will you require four-wheel drive?**
- If you will be driving across pastures and working in snow or mud, four-wheel drive is essential.

# Tractor Wish List Worksheet

Width of gates, pens, aisles or doors _____

_____

Size/Category _____

_____

PTO horsepower _____

_____

Transmission preference _____

_____

Two-wheel or four-wheel drive _____

_____

Bar hitch _____

_____

Three-point hitch category_____

_____

Live PTO _____

_____

Hydraulics, rear _____

_____

Hydraulics, additional _____

_____

Tire type _____

_____

ROPS _____

_____

Canopy_____

_____

Cab_____

_____

Loader _____

_____

Color (Make)_____

_____

Budget constraints _____

_____

New or used _____

_____

# 3 Buying a Tractor

Once you have made your tractor wish list, you are ready to start shopping. Generally you can buy new and used tractors and farm equipment one of three ways — through a dealer, by private treaty, and at auction. All three of these options are available in person or via the Internet.

If you are set on getting a brand-name tractor with low hours and great looks, realize that such tractors are in high demand and that you may have to pay a little over book value to purchase one. Don't expect a used tractor to be ready to roll. Often the reason it is for sale is because it requires some repairs.

# Buying from a Dealer

Working with a reputable dealer in person is ideal, because you will receive experienced advice and have a much better chance of ending up with what you really need. A good dealer will ask you to specify what type of work you plan to do with your tractor, the size of your acreage, and how much you expect to use the tractor. He will also help you project your future needs.

Most dealers offer warranties on new equipment, and some offer a limited warranty on reconditioned equipment. Many of them offer parts and repair service and will provide advice and answer questions after purchase. A dealer may also offer a good deal on financing from the manufacturer.

Ask your dealer to turn your tractor wish list into a written bid. If the tractor he suggests is not on the lot, make sure that the bid includes a delivery date. Talk about financing in detail before you clinch the deal. Use the criteria outlined in this and the previous chapter to help you in your transaction with a dealer.

For novices, buying through a reputable dealer is usually the safest route. But if you already have experience with machinery and equipment, buying privately or at auction can be a viable option. However, unless you have driven, hitched, and pulled a fair bit, get a professional tractor mechanic's opinion before you actually buy a private tractor or bid on one at an auction.

**A good dealer offers selection, support, and service, so is often the safest buying option.**

**Auctions can be a great place to purchase used equipment but if you are inexperienced, take along a mechanic or knowledgeable friend.**

# Buying at Auction

Buying at auction can be intimidating because it's a situation where a novice can get into trouble. The more you know, the better deals you will find. That's why you should only consider purchasing at auction if you have knowledge and experience or if you take someone along who does.

Most auctions have a preview day the day before the auction. It is in a seller's best interest to get his tractor and implements to the sale yard a day or two ahead of the sale to allow buyers to view them. On preview day, get a copy of the conditions of the sale so that you understand all the terms completely. Except on rare occasions, items at auctions are sold "as is," and it is a case of "buyer beware." Each tractor or implement is assigned a sale lot number, so make a note of the lot numbers of the items that interest you. At most auctions, items are presented in numerical order.

Some sellers put a reserve selling price on their items, which means that if the item is not bid up to at least that price, the item will not be sold. The seller generally pays the sales commission to the auction yard but some auctions charge a buyer's commission to reduce the fees to sellers, so be sure you know exactly what you are responsible for paying.

Buyers must register at the auction office prior to the start of the sale to receive their buyer's number and sign an agreement to the terms and methods of payment. In some cases you may be required to leave a deposit. You will need your buyer's number when bidding, as the auctioneer will ask you to show it. When you want to bid on an item and you are sure of the price you are offering, just raise your hand and the auctioneer will acknowledge your bid. If you are not sure of the amount that is being asked, request the auctioneer or one of his bid takers to repeat the amount for you.

Once you have bid on all the items you want, you will have to pay for and remove the equipment you won according to the terms of the auction yard. Along with your receipt for payment, you should receive any other paperwork pertinent to your purchase, such as warranty transfer, operator's manual, and maintenance log, if there is one. After payment, show your receipt to the auction yardman, who will release the tractor or equipment to you. If you have your own trailer for transport, you must load and secure your tractor and remove it from the grounds as soon as possible, usually during a stipulated time period. Some auctions have hauling services available.

Auctions can be a great deal if you know your stuff and have a trailer to haul away your purchase. For many people, however, auctions can be risky, and purchases can wind up costing quite a bit more than buying through a dealer. Hidden costs include time off work, travel time, fuel, and the delivery cost or trailer rental for delivery and future repairs.

# Buying from a Private Seller

When you buy a tractor from a private party, the interaction can develop into more of a relationship, so it is best to start off on the right foot and be friendly, while at the same time keeping your eyes and ears open. If you are new to tractors, one thing you don't want to say to a seller is, "I'll take your word for it," or "What would you do?" Although there are many honest, ethical people out there, others are selling a tractor because there is something wrong that they don't want to take the time and money to fix. Some

A for-sale-by-owner tractor could be a sweet, one-owner, functional tractor or an unsuitable, outdated, or broken down vehicle.

people intentionally talk up the condition and ability of a tractor, or even give you false statistics, while others unknowingly give you the wrong information. Another scenario is the person who knows nothing about tractors, but is selling one that has been sitting for some time after the owner's death or some other circumstance. The seller might know nothing about the history of the tractor, so you will need to do your homework or bring someone who is a good "Tractor Sherlock."

Although it may take longer, it is best to scour the classifieds for tractors for sale than it is to place a "Wanted: Tractor" ad yourself. Think of the time you spend as a free education. You'll gather all sorts of interesting information about tractors and things agricultural and otherwise as you chat with tractor sellers. The ideal seller is someone planning to upgrade from a tractor that is fully operational and currently in use.

If you see a tractor listed in the paper or along the road with a "For Sale" sign on it, first find out why the tractor is being sold. Is it mechanically functional? If not, you'll need to determine repair needs and what the cost would be for parts and labor. An important issue is whether parts are readily available. See page 27 for specific things to look for when buying a used tractor. If you find a used tractor for sale privately or at auction and ask a tractor dealer or mechanic to evaluate it for you, be prepared to pay an appraisal or consultation fee for the professional service.

## Closing the Deal

Before you bid at auction or sign an agreement with a dealer or private party, check the hours on the tractor and its book value. Dealers can usually provide you with the book value or you can find it on the Internet. Once you determine the amount that is appropriate to pay for a certain tractor, and a price that works with your budget, stick with it.

If a tractor seems like a good deal, determine if it truly is, and if so, buy it quickly! If a deal looks too good to be true, however, you might want to check with your local authorities to see if the tractor is mortgaged or stolen before closing the deal.

## Buying Mistakes

Some of the most common mistakes made in purchasing a tractor include being excited but uninformed, using price alone as a buying criteria, buying a fixer-upper, and buying too small. A single feature on a tractor could make it unsuitable for your use. If you are not aware of these types of things, you could end up with a tractor that seemed 100 percent right for you but turns out to be 10 percent wrong — and that 10 percent makes the tractor unusable. For example, tractors with narrow front ends (a single front tire or front tires 6–8 inches apart) are highly maneuverable but very unstable. They have a reputation for tipping over, one of the main causes of injury and death on the farm. A tractor with a narrow front end presents a risk that should be avoided in any situation, but particularly on hilly or mountainous terrain.

Other poor purchases would be a tractor with no power take-off (PTO), no three-point hitch (referred to as a "bareback" tractor), or no rear hydraulics. These three features are musts.

A tractor with a narrow front is an anachronistic deal breaker. It is unstable, unsafe, and difficult to resell.

You can also make a mistake if you approach the matter of buying a tractor solely with price in mind. You may pay a low initial price but end up adding, replacing, or repairing many costly items. Do not consider purchasing a fix-up project unless you are very experienced with tractor repair and parts availability.

When you do buy, make sure you are comparing apples to apples and oranges to oranges. For example, a listed tractor may state that it has a PTO, but not all PTOs are equal. (See page 12 for the difference between a live and a two-stage PTO.) Similarly, the type of transmission you choose will affect the ease and efficiency of your work.

Within reason, it is usually better to buy a more powerful tractor than your current needs require, because you will almost always find more demanding work to perform with your machine. It is a costly mistake to buy a small tractor and then try to accomplish work more suitable for a big tractor. The mechanical workings of the small tractor simply will not hold up, and you will be faced with extensive repair bills. No single tractor will do all jobs for you perfectly, but try to choose the type of tractor that is best suited for your needs.

# New or Used

We all dream of having a shiny new tractor, but just like cars and trucks, new ones are costly. Tractors do retain their value better than many cars and trucks and, if well taken care of, certain models not only hold their value, but actually appreciate as they age. Whether you buy new or used will depend on your budget, how long you plan to keep the tractor, and your uses for it. If you are considering buying used, plan to take more time finding your perfect tractor. Although people are always selling, the selection at any given time is usually much smaller in the used tractor market than it is in the new tractor market.

Try to focus on tractors less than five years old with fewer than 1000 engine hours. Whether domestic or foreign, choose a tractor that complies with U.S. safety regulations. Gray market tractors are imported tractors that may not be OSHA compliant. Most often, gray market tractors are used compact diesel tractors, usually from Japan. They cost substantially less than an equivalent domestic tractor.

Avoid tractors that have homemade upgrades on them unless you are experienced enough to evaluate them properly. No matter if you are buying new or used, you've heard it once already and you'll hear it again: Take your time.

# Buying a Used Tractor

Since a new tractor should be in tip-top shape and covered by warranty, some of the criteria in this section are not necessary for a new tractor purchase, but you can customize the checklists to suit your purpose. When you are buying a used tractor, however, you should consider each of the factors in the following sections and checklists.

Whether you are buying a used tractor from a dealer, a private party, or at auction, try to find out as much information as you can. Be observant, but don't approach a seller with distrust. There is so much you want to know about a tractor that it helps if you break the buying process down into the following four stages and progress through them for each tractor you see to narrow the field: essentials, history, physical inspection, and test-drive.

## ESSENTIALS

Whether you are looking at a tractor on a lot, reading an auction flyer, or responding to a classified ad,

**New or used? While the old utility gas tractor on the left has character and might be perfectly functional, the new one on the right has a diesel engine and the latest safety features.**

## Tractor-buying Kit

- ❏ Coveralls
- ❏ Gloves
- ❏ Rags
- ❏ Clipboard with checklists and pen
- ❏ Camera
- ❏ Large sheets of cardboard (dismantle a large appliance box)
- ❏ Grease gun
- ❏ Coin
- ❏ Flashlight
- ❏ Adjustable wrench

first find out if the tractor has the essentials. Below is a checklist of the major features that you should consider for a horse-farm tractor. Tailor the list to suit your needs and then make a copy for each tractor you are considering purchasing. You can get most of this information without actually seeing the tractor. At this stage, you just want to establish if the tractor is a prospect.

## HISTORY

When you find a tractor that has the essentials on the preliminary checklist, you might want to ask the seller some additional questions over the telephone. This information will help you determine if you want to go see the tractor and take a test-drive. Here you are trying to get a feel for what the tractor has been used for and its condition.

## Used Tractor Worksheet

**ESSENTIALS**

Tractor make and model _____

_____

Owner's name and phone _____

_____

_____

Does it have:

❏ Diesel engine or gas _____

Horsepower _____

❏ Live PTO

❏ Drawbar

❏ Three-point hitch

Category _____

❏ Rear hydraulics

❏ 4WD

❏ Loader

❏ ROPS with seat belt and certification plate

❏ PTO shield

❏ Fenders

Notes: This tractor needs the following repairs to make it functional and bring it up to safety code: (e.g., head gasket, brakes, ROPS upgrade) _____

_____

_____

_____

**HISTORY**

Ask the seller the following questions:

How old is the tractor? _____

How long have you had it? _____

What have you used it for? _____

_____

_____

_____

How many hours has it been run? _____

Does it start and work now? _____

_____

Why are you selling it? _____

_____

_____

_____

Has it had any accidents or repairs? _____

_____

_____

_____

What are the tractor's strong points? _____

_____

_____

What are its weak points? _____

_____

_____

## PHYSICAL INSPECTION

If the tractor still sounds promising after the preliminary and secondary checks, you should take a look at it. During your visit, you can ask the seller for the following information, but you also want to evaluate the tractor yourself and record your own observations. Don't be in a rush and don't be rushed. Take out your inspection kit, lay the cardboard under the tractor to check for leaks, and start taking notes.

### Overall Appearance

What is the overall appearance of the tractor? Is it tidy or cobbled together with bits of tape and wire and rags tied on here and there? Often these temporary repairs are Band-Aids that cover up worse problems. If the tractor has had a recent fresh coat of paint, that could be a good thing (a conscientious seller) or a not-so-good omen. Paint can be an attempt to cover up damage, rust, and welds. If a cheap and quick paint job was done, with unprofessional sandblasting, sand could have gotten into places it doesn't belong.

Has the tractor been stored indoors or outdoors? Rust, weather-checked tires, a brittle seat, cracked steering wheel, oxidized paint, and fogged, cracked, or water-filled gauges are all telltale signs that a tractor has been stored outdoors. Besides the obvious external damage, these symptoms indicate that moisture has probably seeped into other places as well, such as bearings, hydraulics, oil, and the fuel system. Once rust has started, it is difficult, if not impossible, to stop. Pay particular attention to the floor, especially if there is a floor mat, as moisture could have been trapped under it over the years and there could be a black hole there ready to implode.

### Identification and Records

Is there a products identification number (PIN) plate? On the plate, there should be a model number and serial number. You might find it fastened to the engine, under the seat, or on the side of the dashboard. Numbers that are stamped into engine castings can also be identification numbers, but

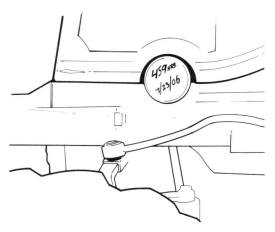

The notations on this oil filter indicated the date and engine hours when the filter was last changed.

embossed (raised) numbers on castings are manufacturer instruction numbers and do not pertain to identification.

Although tractors do not have titles like cars and trucks do, the seller should have a bill of sale that indicates when he purchased the tractor and other information. You might want to ask the seller to produce a bill of sale or other ownership documents if you have doubts about the legal ownership.

Does the seller have a maintenance schedule and log? Ask if he has kept track of major repairs and if you can see the log. While you are at it, you might ask to see the operator's manual.

### Structure

Is the tractor structurally sound? Look at the frame and inspect all steel and cast components for obvious cracks or hairline cracks. Keep an eye out for signs of welded repairs. Look for unevenness, cracking at the weld, or improper alignment of the parts. Ask the seller, "What happened here?"

The rollover protective structure (ROPS) should be certified, not homemade. Look for a plate that tells you so. Any certified ROPS will have a seatbelt. Note that cabs themselves do not necessarily meet ROPS approval. They may crumple in a rollover. If a tractor needs to be retrofitted with a ROPS, it will probably cost more than $1500.

If the tractor does have a cab, check to see if all

**The condition of the air filter can indicate the amount and quality of air getting into the engine.**

the windows and doors work. What kind of seats does it have? Metal pan seats can be very uncomfortable. Look for a seat with a good cushion and supportive back.

### Filters

Are the filters clean and in good shape? The filters you should identify and inspect include the oil filter, fuel sediment filter, hydraulic filter, and air filter. When you look at the filters, can you read them or are they covered with a thick layer of crud? The air filter is particularly important — it should not be crumpled or damaged or filthy. The health of the engine depends greatly on the quality of the air and fluids it receives. A clean new filter does not necessarily indicate that the tractor has received regular servicing, but there's a better chance that it did than a tractor whose filter you can barely find.

### Fluid Levels

Are the fluid reservoirs full? Grab your rag and check the following fluid levels: oil, transmission, power steering, differential, hydraulics, and antifreeze (radiator).

What does the oil look like? Pull the dipstick, keeping an eye out for presence of water or foam, which would indicate a serious problem. If the oil looks whitish at the top of the dipstick, it indicates a water/oil emulsion. Check again after you've taken a

test-drive. By the way, black oil is perfectly normal for diesels.

Transmission fluid should be pink, not brown, and should smell fresh (almost sweet), not burnt.

When removing the radiator cap, note if there is a white creamy substance on the cap. This might indicate gas or oil in the system. Are there signs of overheating on the neck of the radiator and along the top? If so, there might be problems with the thermostat, water pump, or more. Look at the external radiator fins for damage. The more dings and dents, the more likely you will have to replace the radiator.

### Battery

How does the battery look? The battery should be firmly anchored, with posts clean and protected. If the caps are removable, look inside. The water level should cover the plates. If it looks dry, the battery is likely shot. You can expect to change a tractor battery about as often as you do a truck battery. Replacing a battery is a reasonable repair to make on a used tractor.

### Belts and Hoses

Are the belts and hoses in good working order? While you are under the hood looking at fluid levels and the battery, check the belts and pulleys carefully. Look for cracks or glazing on the belts: The former

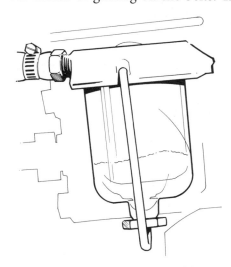

**It is not a good sign if the sediment bowl contains crud or water.**

indicates it is past time to replace it; the latter shows that the belt is worn, slipping, and inefficient.

A good hose is smooth and firm. Squeeze hoses to reveal cracks that could be a few minutes away from bursting. Also feel the hoses for excessive sponginess, which indicates the integrity of the hose is compromised.

## Tires

On most used tractors, the tires are quite old and you can expect to see some weather checking (cracking) from the deteriorating effects of sun, mud, and water. If the tires are showing deep fissures, however, beware — they could come apart at any time. Generally speaking, tread depth is not as important on tractor tires as it is on your car or truck. However, look for lugs with at least 50 percent of their

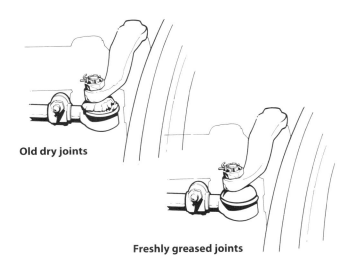

**Old dry joints**

**Freshly greased joints**

tread depth left. New 28-inch tires have 1-inch tread depth, and new 38-inch tires have 1½-inch tread depth. Slip a dime (which is ¾-inch diameter) in the tread to measure.

## Grease Zerks and Wheel Bearings

Does it look like all moving parts have been greased regularly? Look for excess grease around moving parts. Remove the grease fitting and look inside for a clean supply of grease. If the well is empty or the grease is dirty and clumped, the tractor has not received proper lubrication. Do the zerks still take grease? Or are they clogged with dirt and the joints "frozen"? Wipe off the zerks and try to add grease to them with a grease gun. You should see old grease being forced out of the joint — not water and not just the new grease. Manually or with tractor power, manipulate joints to see if they move freely without being sloppy. Too much play indicates excessive wear and tear.

Grease is supposed to be *in* the wheel bearings, not on the outside, so if you see a collection of grease on the outside, it can mean the bearings and/or seals need replacing. Wheels with worn bearings have play in them, which you can feel when you rock the wheel from side to side.

**Above: The deep cracks of weather checking and age make these tires unsafe. Below: The condition of the tread can affect how much traction a tractor has.**

## Weights

Are there wheel weights or suitcase weights on the tractor? Note the amount and placement of the weights. Ask the seller if the weights were added to balance the tractor for a particular implement or task. Are the tires fluid-filled to add weight? Find out how much and what type of fluid is in the tires and how long it has been there. Check for signs of rust on the wheels that could indicate leakage.

## Miscellaneous

Do the headlights, taillights, and turn signals work? Does the horn work? Are the controls clearly marked as to what they do and how to use them? Are the safety and warning decals still in place and legible? When you climb up to the driver's platform, are there adequate steps to help you mount? Are there handholds or other safe places you can grab onto as you climb?

## Physical Inspection Checklist

### OVERALL APPEARANCE

Stored indoors or outdoors _____

Rust _____

Paint _____

Steering wheel _____

Seat _____

Gauges _____

### IDENTIFICATION AND RECORDS

Model _____

Serial number _____

Bill of sale _____

Maintenance schedule and log _____

Repair records _____

Operator's manual _____

### STRUCTURE

ROPS / seatbelt _____

Certification plate _____

Cab _____

Doors _____

Windows _____

Floor _____

Mounting steps / handholds _____

### FILTERS

Oil _____

Fuel _____

Hydraulic _____

Air _____

### FLUID LEVELS

Oil _____

Transmission _____

Differential _____

Power steering _____

Hydraulics _____

Antifreeze _____

Battery _____

Grease zerks and wheel bearings _____

### MISCELLANEOUS

Belts and hoses _____

Tires _____

Weights _____

Operating lights _____

Horn _____

Marked controls _____

Safety warning stickers _____

## TEST-DRIVE

If a prospective tractor passes the first three stages of evaluation, the next step is to run the tractor and take a test-drive. Plan to spend at least half an hour on this step. Just like buying a horse, ask the seller to demonstrate things first. If you are a novice tractor driver, you might ask for a brief orientation of the controls. Note that not all sellers will allow buyers to test-drive a tractor due to liability concerns. In that case, ask the seller to demonstrate the tractor in action. One way or another, evaluate each item on the accompanying checklist.

The first portion of the test is stationary; the second part is the actual test-drive.

### Cold Start

First make sure the engine is cold. Turn the key on. Did the oil pressure light go on when the key was turned on? Then start the tractor. Did the oil pressure light and all other engine indicator lights go off? Did it start easily? A tractor that starts cold probably has a good battery, compression, and ignition, and a functioning fuel system. If it doesn't start easily, it may be a good tractor in need of a battery or tune-up or it may have serious problems.

If a tractor is basically in good shape, a conscientious seller should have taken care of minor things (battery, plugs, flat tires) so that the tractor starts and is drivable for a potential buyer. If there are relatively simple things that prevent you from running the tractor or driving it, the reason could be a red flag that the seller is trying to cover up a worse problem.

If you can't start it, it will be hard to evaluate other things, so be pretty ruthless here. If a tractor is already running when you come to look at it, make all of your other inspections and then if still interested, make an appointment to come back later or on another day so you can start it cold.

### Oil Pressure

What is the oil pressure reading on the gauge on start-up and when hot (after running for 30 minutes or so)? If the gauge does not work, have a mechanic test the oil pressure for you. Oil pressure is higher when the oil is cold and thick, and drops as the oil warms up and thins down. A gas-engine tractor should carry at least 15 pounds of oil pressure when warm; a diesel, 20 pounds.

### Charging System

To thoroughly test the charging system, turn on the headlights for a couple of minutes before you even start the tractor. Now start the tractor and note if the headlights get brighter (a good sign). Then turn the lights off. When the engine is idling, there should be a slight charge on the ammeter. Turn the lights on again and the gauge should note a slight discharge. When the tractor is at running speed, the ammeter needle should be dead center, neither charging nor discharging.

### Smoke

Diesel tractors normally puff a little black smoke on start-up, but soon should run clean. Blue smoke could mean problems with rings, pistons, or valve guides. Continuous white or black smoke could indicate the need for a tune-up (carburetion or ignition). If all else about the tractor checks out, you could bring your mechanic by to help you evaluate the source of the smoke.

### Noise

What kind of noises do you hear? A running engine should sound steady, smooth, and rhythmic. Ticking might mean a valve adjustment is needed. Loud knocking, clunks, or thuds are usually serious, involving the crankshaft, bearings, or pistons. Listen at an idle and at several higher rpms.

### Belts and Pulleys

Remove the access panels and look for looseness or wobbling while the engine is running.

### Leaks

Is there leakage in the middle of the tractor? A faulty

**White smoke could indicate the need for a tune-up, while blue smoke might mean more serious problems.**

front or rear main bearing oil seal can leak oil on the clutch and ruin the clutch lining. If the rear main seal needs to be replaced, the tractor will usually need to be split in two; this is a costly repair. Your cardboard should still be under the tractor. Look for spots on it and check underneath the tractor near the oil pan for leaks.

Do you see oil leakage around the head gaskets? The gaskets might need to be replaced (not really a big deal) or the heads might be warped or cracked (a very big deal). There might be signs that the head has been recently removed — if so, you might see a variation in paint color between the head and the block. This could mean that the heads have recently been replaced or that a valve was bad. It would be nice to know.

Also note whether there are fuel leaks (usually detectable by smell), coolant leaks, hydraulic oil leaks at the couplings or from damaged hoses, or oil leaks at the PTO seal or the four-wheel-drive seals.

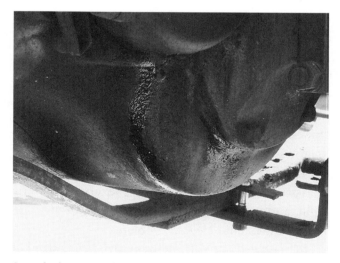

**Some leaks you can live with, others must be repaired. Ask for a mechanic's opinion before you buy.**

## Transmission, Clutch, and Drive

Shift through every gear in each range while feeling for ease of shifting and listening for grinding or thunks.

If there is a clutch, choose an appropriate range and gear and test it for slipping. There are two ways to test the clutch. First, with the tractor running, push in the clutch and put the tractor in gear. If it grinds, the clutch could be on its way out. Next, put the tractor in low gear and with your foot on the brakes, slowly let out the clutch. The engine should lug down and die. If the clutch slips and the engine keeps running, the clutch is shot or close to it.

If you are testing an automatic transmission, put the tractor in a low gear. Then apply the brake and hold it while applying the accelerator. The tractor should try to move and then stall. If it doesn't stall, the transmission might have problems. Keep your foot on the brake and go through the gears from park to reverse, from reverse to forward, from park to forward and listen for any loud clunking that could signal transmission trouble.

As you drive, accelerate and decelerate and listen again for clunking in between; this could mean axle or U-joint problems.

## Power Steering and Turning

When the tractor is running and parked, there should be little or no play in the steering. Play is an undesirable looseness that allows you to move the steering wheel to the left and right without the tires moving. An inch is okay; more is suspect.

Turn the wheel all the way to the left, then to the right, and listen for squealing of the power-steering pump, which could signal problems.

When the tractor's front tires are pointing straight ahead, the steering wheel should be centered. If not, it could indicate alignment problems.

Keeping an even speed throughout, drive the tractor on fairly level ground, make a turn and then let go of the steering wheel (or at least lighten your grip so you can feel what happens). The tractor should straighten out on its own.

As you drive, note the turning radius. Although this can be provided to you as a number from the manufacturer, there is nothing like firsthand experience to see if it will work for you.

Do the wheels wobble when driving straight? Ask a friend to observe the tractor as you drive toward and away from him. Wobbling can indicate loose lugs, alignment problems, or a bent frame.

## Four-wheel Drive

Does the four-wheel drive work? The best way to test this is to climb a small incline or turn tight circles to see if the front tires are pulling. Is the 4WD easy to engage and disengage?

## PTO, Three-point Hitch, and Hydraulics

Does the PTO work in all gears? If possible, attach a three-point PTO implement so you can continue your evaluation. One arm of the three-point hitch has a crank to adjust it up and down. Does the crank turn freely in both directions?

Do the hydraulics work without leakage or letdown? Manipulate the hydraulics to the loader or implement, paying attention to whether they operate smoothly or chatter. Chattering indicates the pump is worn and could be ready to fail. Lift an implement with the hydraulics and leave it up for a few minutes to see if there is any loss of lift power. If the implement starts lowering on its own, the hydraulics require attention. Even if you turn the tractor off, the implements should stay up.

Were the hydraulic couplers capped when not in use? If not, dirt and water have probably entered the hydraulic system. Rusty balls in the couplers indicate poor maintenance.

Inspect (wiggle) hoses and fittings for signs of wear and leakage. Are the hoses sound, or are they chafed or split or melted in spots? Some sellers cover damaged hydraulic hoses with sleeves, hiding hoses with exposed metal strands that are in a potentially dangerous condition — capable of bursting at any moment. If possible, detach and reattach the hoses to get a feel for the ease of operation.

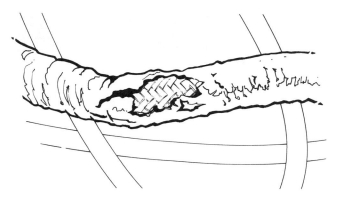

A hydraulic hose with significant damage to the rubber sheath will need replacing.

Hose sleeves can be used to prevent damage to hydraulic hoses or to hide existing damage.

## Loader

If the tractor has a front-end loader, it might have a separate hydraulic pump for the loader; test it by raising the bucket, lowering the bucket, and tilting the bucket forward and back. This will indicate the integrity of the loader's hydraulics as well as the function of the joystick and other controls related to the loader.

## Running Hot

Does the tractor run well when hot? Run the engine for half an hour while you complete the other tests and keep an eye out for oil, fuel, or coolant leaks. Note if the oil pressure drops significantly when the engine is hot, or if there is a change in smoking or sounds, or any other signs that cause concern. You will never really know how a tractor will perform after two hours of mowing on a hot day in July until

you are actually mowing, but a test-drive can highlight obvious areas of concern. After a half hour of running, is the temperature normal or does the tractor run too hot or too cold? If so, there might be a problem with the thermostat, the water pump, or the temperature gauge.

## Brakes

Tractors have one brake for each rear wheel. When testing, check one brake at a time. In low gear move ahead slowly, depress the left brake pedal, and turn the tractor to the left. The left rear wheel should not turn, and the tractor should rotate around it. Do the same with the right brake, turning to the right. Compare the response between the right and left brakes. Then lock the brake pedals together (with the hinged bracket that is on one of the pedals), which is how you usually will use them. They should lock together easily.

Find a level spot and drive the tractor straight. Then depress the locked-together brake pedals and see if the tractor pulls to the right or the left or if it stops straight. Make sure the parking brake works by setting it with the tractor parked in neutral and on a slight incline.

## Blow By

When you have finished the test-drive and are ready to park the tractor, do so on a level spot where you can place your cardboard under it once again to test for leaks. Significant leaks should be evident within five minutes. After you have turned off the tractor, remove the oil cap and look for "blow by" which is exhaust from the cylinders that has leaked into the crankcase. It is not a good thing to see exhaust vapor rising from the oil spout.

Check the oil dipstick again, noting the presence of foam or an odd color.

## Hot Start

After the tractor sits for a few minutes, try starting it hot. If a hot tractor fails to start, there could be a problem with the carburetor or radiator.

## Test-drive Checklist

*COLD START:*

**TURN THE KEY.**

Oil pressure light on? _____

**START THE TRACTOR.**

All engine lights off? _____

Oil pressure reading _____

Charging system _____

Smoke_____

Noise _____

Belts and pulleys _____

**LEAKS**

Fuel (gas or oil) _____

Radiator (coolant) _____

Head gaskets (oil or coolant) _____

Hydraulics_____

Rear main seal _____

PTO seal _____

4WD seal _____

**TEST-DRIVE**

Transmission _____

Clutch_____

Power steering and turning _____

Four-wheel drive _____

PTO_____

Three-point hitch_____

Hydraulics_____

Loader _____

Brakes_____

**NOTE:**

Temperature gauge reading_____

Turning radius _____

Wheel wobble? _____

**TURN OFF AND CHECK.**

Leaks _____

Blow by _____

Dipstick: level, color _____

*HOT START:*

Starts after sitting for 5 minutes? _____

# Tractor Implements

Before you purchase a tractor, determine which tasks you will be doing and check the availability of implements to fit the tractor you have in mind. Today you can find almost any implements to fit the smallest garden tractor to the most mammoth farm tractor.

Note that many implements are also available for all-purpose vehicles (APVs), which are discussed in the next chapter. Try to divide your tasks into big jobs and small jobs. You might need a full-size spreader and mower for your tractor, but find a weed sprayer and feed wagon handier to operate from your APV.

Avoid a package deal (tractor plus implements) unless it contains exactly what you need, such as a front-end loader. Don't pay extra for an option or an attachment just because the person selling it says it comes with the tractor. When buying used, choose only those items in very good working condition. If in doubt, get professional advice. If you don't need a particular option or implement, ask what the price of the tractor would be without it.

# Loader

A front-end loader is a large bucket mounted on the front of the tractor and used for scooping up and moving quantities of material. On a horse farm, that material is most often manure, but a front-end loader can move gravel, dirt, bedding, and even snow. If your tractor didn't come with a loader, you can add one (made by the same manufacturer or a third party), along with an auxiliary hydraulic system, as long as the loader is specifically designed to work on your tractor. (See chapter 2 for more on hydraulics.)

When choosing a tractor and loader, consider breakout force, which can range from 900 pounds to 9000 pounds, and maximum load lift capacity, which can range from 515 pounds to 5140 pounds. Breakout force is a loader's capacity to break through and lift hard-packed earth.

Consider also maximum lift height, which can range from 75 inches to 169 inches, clearance with the bucket in the dumped position, and the reach of the arms at maximum lift height. All of these figures will help you determine if a tractor has the power and scope to reach where necessary, such as over fences, panels, walls, or deep into sheds.

The range of motion of a bucket and how quickly it moves will contribute to the time required to cycle through a repetitive task such as loading manure onto a spreader. If working rapidly is important, you might want to consider a self-leveling loader, as it will decrease spilling and allow you to keep a quick work pace.

A mid-mount loader frame, which attaches just ahead of the operator, is best. That's because it eliminates the need for bulky bracing at the front of the tractor. Valves and hoses that are mounted at the mid-section instead of the rear of the tractor make hookup easier. Pay special attention to the mounting dimensions, size of the loader in relation to tractor horsepower, and operating clearance. As you shop, check to see what other attachments and different bucket sizes are available that can be interchanged with your standard bucket.

**A front-end loader is essential for handling bedding, manure, gravel, dirt, and snow.**

A pipe-frame trip-bucket loader, usually found on older tractors, can be raised and lowered hydraulically, but the tipping of the bucket itself is an all-or-nothing situation. With their limited control over the angle of the bucket, trip buckets are notorious for either skipping over the top of your intended load, or digging in too deep and tearing up the earth underneath.

Double-action hydraulic buckets are much more versatile than trip buckets, because you can control the position of the bucket with a much finer degree of movement. This allows you to sprinkle the material you are unloading rather than just dumping it in a pile as is the case with a trip bucket. You operate the bucket with a single joystick hydraulic control that is intuitive for many novice operators. All four movements (lifting bucket, lowering bucket, tilting bucket forward and tilting bucket backward) are performed with one lever.

If you will be using several sizes of buckets — perhaps a large one for sawdust and smaller, heavier one for gravel — a quick-attach option for the entire loader and/or the bucket is handy. It makes it simple to detach the bucket if it obscures your view and you want to remove it when not in use, or if you want to replace the bucket with another implement. Some quick-attach designs require tools and aren't as convenient as ones that only require the flip of a handle. Some implements can be added right to the bucket using bolts or chains — this would include such things as a manure fork, bale spear, or pallet fork. Some loaders even offer a quick-attach device that enables you to use a backhoe or posthole digger attachment designed for skid steers.

Keep in mind that you may have to weight your tractor's tires or add ballast elsewhere (on the frame or at the rear) to properly balance your tractor for loader operation. Consider this very carefully, especially if you plan to attach other implements to your loader or carry heavy material in the bucket; doing so positions the weight much farther ahead, out in front of your tractor.

A tractor with a loader should have a grill guard to protect the front of the tractor from material or items falling out of the bucket. If shopping for a used loader, look for signs of stress: bowed arms, bent bucket, welded repair spots, worn-out pins and bushings, and leaking hydraulic cylinders.

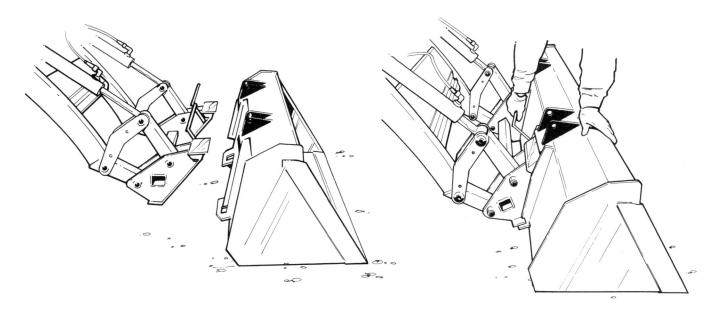

**Quick-attach systems make changing implements easy.**

**You only have to engage two hand levers to lock this bucket in place.**

# Harrow

Harrows, also called drags, break up and smooth out plowed or clumped soil. They are useful for smoothing an arena, round pen, or track after disking; breaking up and spreading manure in pastures; working fertilizer, herbicides, or seed into cropland; and aerating compacted soil.

Some harrows have adjustable tines. You can set them straight down for maximum penetration or angle them backward to break up surface clumps. Some harrows can also be flipped upside down with the tines pointing upward for use as a drag mat. Harrows can be ganged, with one following the other to perform two chores at once. For example, the first one can be set with the tines straight down to dig up the soil while the second one has the tines backward to break up the clumps, or the first one could break up clumps with backward tines while the second one is flipped over to smooth out the surface.

There are five basic types of harrows: English (also called blanket, chain, or flexible tine), spike tooth, spring tooth, rotary, and combination. Each is suitable for specific tasks, so over time, you might find it necessary to purchase more than one type.

The **English harrow** is made of heavy wire stock (⁷⁄₁₆ to ½ inch) that is crisscrossed in a diamond-shaped configuration and may or may not have protrusions called teeth (tines) on the bottom side. The teeth, which make the harrow more aggressive, range from ⁷⁄₁₆ of an inch to ⅝ of an inch in diameter and are from 3½ to 4 inches long. A good English harrow is usually heavy and costly, but does a wonderful job of smoothing rough spots. They are excellent for leveling gravel driveways, spreading manure in a pasture, and aerating the soil without ripping it up. Homemade drags made of chain link fence attempt to simulate the English style, but they lack teeth, and their light weight makes them bounce on top of the soil rather than doing much smoothing or leveling.

Like pull-type discs, some English harrows can be difficult to load, and when you move them by dragging them behind the tractor, they harrow everything behind you. Others attach to the three-point hitch and can be lifted for transporting. Some can be rolled or hung up for convenient storage.

An English harrow connects to the tractor by means of its own drawbar. The harrow's drawbar is a long horizontal bar or pipe that attaches along the entire front of the harrow. You can use a six-foot drawbar for a six-foot harrow, a 12-foot drawbar for two six-foot harrows set side by side or even a 24-foot drawbar for four six-foot harrows.

A **spike-tooth harrow** has a more rigid frame with teeth like old railroad spikes mounted on movable bars that can be adjusted for work or transport. With the teeth in full upright position, you can rip up pastures in the spring; with angled teeth, the harrow works more as a leveler and smoother. Most spike-tooth harrows have replaceable teeth, as the teeth become worn after long or hard use.

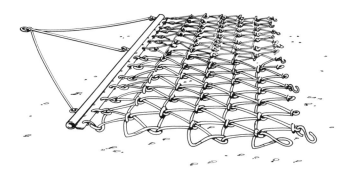

**An English harrow resembles a section of chain-link fence but is much heavier and often has teeth on the bottom side.**

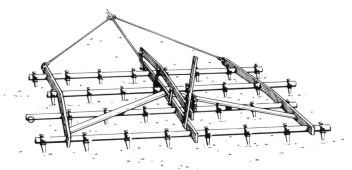

**A spike-tooth harrow's teeth can be adjusted to dig into the ground or to lay flat.**

The **spring-tooth harrow** has crescent-shaped steel tines that allow it to act as a cross between a ripper, a mild plow, and a harrow. Its configuration makes it better at breaking up unplowed ground than actually smoothing out soil. A recent innovation that is a cousin to the spring-tooth harrow is a rake, which is a heavy-duty angled iron frame with 30–50 adjustable spring steel tines. Better suited for landscaping and site prep, it can double as an arena and pasture tool.

The **rotary harrow** is a three-point attachment that has a rigid circular frame with cross braces. It is specially designed to maintain arena footing. Heavy teeth, or tines, on the perimeter of the circle and on the cross braces aerate and smooth the footing as the harrow rotates. The teeth are welded or bolted on; when they wear down, bolted tines can be easily replaced, but welded tines must have new tines welded next to the old ones.

The three-point hitch is adjusted to have the harrow work flat, or tipped to one side, or to the front or back. When adjusted flat, the harrow doesn't rotate, while tipping it to one side slightly causes it to turn as it is pulled forward. Set at its most aggressive tilt, a rotary harrow can dig down six or seven inches, if the soil is not too compacted.

**Combination implements can perform multiple tasks in one pass, saving time and fuel.**

The **combination drag** is several implements in one and could be called a power box rake. It might be an all-in-one chisel plow, leveler (float bar or paddle wheel), scraper box, pulverizer, and roller. It usually is a freestanding implement with two tires. It is generally used for preparing seed beds, but has proven to be ideal for working arena surfaces. In the first pass, stones are collected and directed into windrows for pickup. In the next pass, soil can be conditioned, raked, graded, and leveled. Working in forward or reverse produces different results at variable depths.

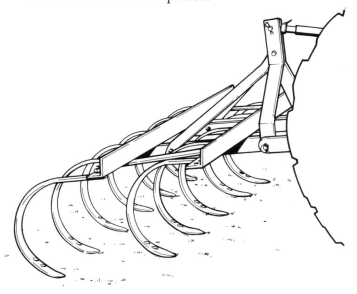

**A spring-tooth harrow is better for breaking up ground than for smoothing it out.**

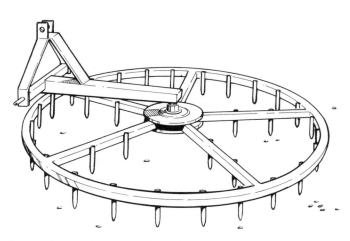

**A rotary harrow rotates as it is pulled by a tractor.**

# Manure Spreader

Manure spreaders are wagons with a mechanical apparatus designed to distribute manure as the tractor is driven through a pasture or field. Smaller spreaders are friction-driven; larger spreaders run off the tractor's power take-off (PTO). If you only spread compost once or twice a year on your fields or pastures and clean the spreader well between uses, a spreader can double as a farm wagon.

Friction-drive spreaders (also called ground-drive spreaders) are ground driven, that is, the power for unloading and spreading is generated by the tires of the spreader rolling on the ground. A ground-drive spreader is usually a simple setup with just two levers: one to control the speed of the apron chain, which moves the load toward the rear of the spreader, and one to activate the beater bars at the back of the spreader. The beater bars break up the manure and fling it into the air. You can drive to your pasture without spreading manure along the way, and then activate the apron chain and beater bar to start spreading.

Since these spreaders don't have a rotating driveline, they are potentially safer than a PTO-driven spreader. The drawback to this type of spreader is that the tow vehicle must be moving for the spreading mechanism to be activated. A friction spreader

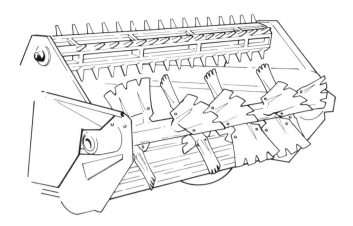

**The beater bars at the rear of a manure spreader break up the manure and fling it.**

can be operated behind a tractor, a pickup, or even a team of horses, though in order to use a spreader with a team, you will have to purchase a special conversion kit.

Spreaders powered by a PTO are usually large, heavy-duty machines suitable for a horse farm or ranch as opposed to a small acreage. The PTO makes them more difficult to hook up, but they have several advantages, one being larger capacity. Another advantage is that because the spreader speed can be controlled separately from the ground speed, the manure can be spread heavily or lightly or even piled in one spot if desired.

**A small friction-drive spreader would be suitable for daily stall cleaning and spreading.**

**A large capacity PTO-driven spreader is ideal for spreading annual compost.**

## SPREADER OPTIONS

When considering spreaders, look for one with up-front controls that you can operate from the tractor seat. Hydraulic controls are more desirable than manual levers. Some spreaders have as many as five apron-chain speeds. There can be up to three sets of beater bars (with ripper teeth) or paddles at the rear of the spreader. Spreaders are available with standard tires or floatation tires for soft ground. Steel wheels without tires are seen on older spreaders.

An end gate is an option that comes in handy if you will be hauling wet loads or if you want to heap the spreader to capacity. End gates are either manual or hydraulic. A front box extension builds up the front of the spreader to allow more heaping and to prevent manure from dropping on the drive mechanism.

Spreader floors are made of ¾-inch tongue-and-groove polyethylene, marine plywood, or steel. Some spreader boxes have recycled high-density plastic liners. A slick surface is preferable because manure doesn't freeze to it and it is easy to clean. Choose a floor material that doesn't warp, bulge, or bow, which could hamper the movement of the conveyor bar.

A slip clutch can protect your chain and beaters. A large frozen clump of manure wedged in the beater can break shear bolts, snap the chain, or damage bars or paddles. With a slip clutch feature, the spreader mechanism stops operating when it gets clogged.

## WHAT SIZE SPREADER

| WHEELBARROW EQUIVALENT | CUBIC FEET SPREADER | # OF STALLS OR PENS | TYPE OF TRACTOR |
|---|---|---|---|
| 2 | 10 | 1 | Garden tractor, ATV, or UTV |
| 5 | 25 | 2–4 | Garden tractor, ATV, or UTV |
| 7 | 35 | 4–6 | Garden tractor, ATV, or UTV |
| 11 | 55 | 5–10 | Compact |
| 15 | 75 | 12–20 | Compact |
| 20 | 100 | More than 20 | Utility or farm |
| 25 | 125 or larger | Commercial stable | Utility or farm |

**If you plan to clean stalls directly into a spreader, be sure it fits in your barn aisle.**

To comply with regulations requiring thin application to reduce contamination from runoff, you can purchase optional equipment to further reduce the conveyor speed. This is especially good if you spread fresh manure or manure without bedding.

### SPREADER CAPACITY

Spreader capacity is measured in cubic feet. You will see two figures in spreader capacity specifications — struck and heaped. Struck refers to a level load and heaped is the mounded capacity. What size spreader you choose will depend on whether you spread manure daily or seasonally. If daily, you may prefer a smaller spreader that can be pulled by a garden tractor, all-terrain vehicle, or utility vehicle, and that is narrow enough to fit in pens and barn aisles or to back into a stall.

If you compost manure (the most environmentally responsible method) and spread it months later as humus, your annual or semi-annual spreading will go more efficiently with a larger capacity spreader, especially if the distance from the pile to the field is far. For five or more horses, figure on a 75-cubic-foot manure spreader or larger. For operations with very large manure-hauling needs, truck-mounted spreaders are available and would be warranted if you need to drive far on highways to spread.

As an example, we have five horses. We collect manure every day, compost it, and spread the humus once a year. We have an International Harvester 540 PTO-driven manure spreader that has a 90-cubic-foot capacity struck and 135-cubic-foot capacity heaped. It takes seven tractor bucket loads to fill the spreader to heaped. We have 70 acres divided into nine pastures. We spread approximately 10–15 manure-spreader loads of humus on our pastures per year, rotating between pastures from year to year.

# Mower

Mowers are designed to cut hay, weeds, or brush. Models are ranked according to the diameter of material they can cut, from about one to four inches. Those at the low end are for turf and pasture, while those at the high end, sometimes called "brush hogs," can chop and shred heavy brush and small trees.

They come in various styles, with rotary, sickle bar, and disc being the most commonly used on horse farms. If your land is uneven, a rotary mower would probably be most appropriate. For cutting hay, a sickle bar or disc mower would be best. If you have brush or heavy weeds that need to be removed, consider a brush hog, which is a heavy-duty rotary cutter/shredder.

Rotary mowers are the most popular and are available as either a pull-type with a self-contained engine, a pull-type powered by a PTO, or a three-

**A heavy-duty rotary mower with large capacity gearbox means cooler mowing during those hot July days.**

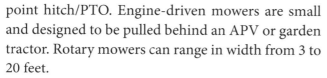

**This pull-behind rotary mower has a self-contained gas engine.**

**For large mowing jobs, flex wing mowers provide broad coverage and the ability to adapt to varied terrain.**

point hitch/PTO. Engine-driven mowers are small and designed to be pulled behind an APV or garden tractor. Rotary mowers can range in width from 3 to 20 feet.

The mower can be raised up so the blades can be used at any desired mowing height up to about 13 inches. This is handy when you want to lop off the tops of broad-leaf weeds but not mow the grass. Pull-type mowers can be raised or lowered with a manual crank or a hydraulic assist cylinder. Three-point hitch mowers are raised and lowered with the three-point hitch.

Rotary mowers have from one to four tail wheels, depending on the size and style of the mower. Tail wheels help support and balance the mower to minimize scalping (cutting into the earth) on uneven terrain. Tail wheels swivel 360 degrees and their height is adjustable.

## ROTARY MOWER OPTIONS

Most rotary mowers have rubber or metal deflector shields to prevent rocks or other debris from becoming airborne missiles. Chain guards are optional for the front and rear of the mower. They offer further protection and don't get hung up on uneven terrain or affect airflow as much as solid guards do.

Another good feature to look for is shear pin

driveline and slip clutch protection, so that if the mower does happen to contact a large rock or stump, the safety mechanisms stop the blades and prevent costly damage to the driveline, gearbox, or blades.

A valuable protective feature, which may or may not be an option, is a type of skid plate called a pan guard that helps the mower glide over obstructions without damaging the blades or drive mechanism. Similarly replaceable, and in some cases adjustable, skid shoes can help prevent damage to the side skirts when hitting rocks.

The gearbox will vary in diameter of input and output shaft, and size of roller bearings, depending on how powerful a mower you require. Oil capacity can affect running time. A larger capacity oil reservoir generally means a greater cooling capacity so you can mow longer in hot weather.

Grooming rotary mowers have four wheels, so there is no soil scalping even on uneven terrain. All edges of such a mower stay a uniform distance from the soil at all times. Grooming mowers can be ganged together in groups to mow large areas if your tractor has the horsepower to run them. You can also purchase already ganged mowers that are called flex-wing or flex-deck mowers. The sections articulate and conform to uneven land and fold up for transport.

A sickle bar mower cuts clean but can clog easily.

This disc mower is suitable for haying a large field.

Most mowers are pulled behind a tractor, but some will attach to the front. Mid-mount mowers work with small tractors that will be used in confined areas where pulling a mower behind would be difficult.

## OTHER TYPES OF MOWERS

Sickle bar mowers have a bar of cutting blades that is pulled back and forth by a pitman arm across a fixed bar of cutting blades. These mowers are best suited for flat ground. The bars range in length from six to eight feet. They can be used for trimming banks as well as mowing fields. These mowers cut more cleanly than rotary mowers, but they don't have as variable a mowing height range as rotary mowers. They also cost more and tend to clog.

Disc mowers are a combination sickle bar and rotary mower. They have similar configuration and height adjustment range as a sickle bar mower does. The design of the cutting mechanism results in less plugging than sickle bars and offers higher ground speed than a rotary mower.

A flail mower knocks down high growth with heavy-duty cutting blades that have high-speed knives or flails attached to a spinning drum, making it suitable for coarse plants. It doesn't clog so it is good for mowing along roads, field edges, or heavily weeded areas.

## Mower Blades

Some mowers can use different types of blades. The blade you choose will depend on the kind of mowing you do. Most blades can be sharpened and/or replaced.

- Low lift: Aerodynamic design creates minimal updraft; good in sandy or loose soil
- Medium lift: Average blade good for normal use
- High lift: Strong suction makes plants stand up; requires greater horsepower to operate in dense stands; not for use in sandy soil
- Mulching blade: Irregular edges chop and shred materials as they cut

# Blade

A blade is a moldboard with a cutting edge that is used for pulling or rearranging dirt, manure, bedding, or snow. The moldboard is curved to minimize drag and to roll dirt or snow off the blade. A blade can be rear-mounted (most common), belly-mounted (middle), or front-mounted, although a front-mounted blade is more commonly called a plow. A hydraulically operated front-mounted plow comes in handy in areas with high snowfall, especially for clearing long lanes or roadways to barns and other buildings. Most plows have adjustable skid shoes at the bottom edge to keep from digging into the road.

The typical horse-farm blade is a five- to eight-foot wide, rear-mounted, three-point blade, although blades are available up to 14 feet wide. A blade can be used straight, at an angle, or swiveled completely around (by removing a pin) so that it can be used to push backward. The three-point adjustment can keep the blade level or tilt it to one side or the other. Smaller blades have all manual adjustments; some larger blades have hydraulic adjustments. Depending on the setup, adjustments are made manually — either by a rope attachment from your seat or by getting off the vehicle and moving the blade by hand — or hydraulically from the driver's seat.

A blade is handy for scraping pens, leveling driveways, and moving light snow. For moving heavy snow, a loader or plow works better. A V-plow will enable you to make an initial path through deep snow more easily than a straight plow. Another snow removal option is a snowblower/thrower that mounts like a plow at the front of the tractor to gather snow and blow it into a specified direction through a discharge chute. (See page 56 for more on snowblowers.)

A belly blade (usually six to eight feet wide) is mounted just below the operator at the midpoint of the tractor where the operator can easily see what he or she is doing. There is no need to constantly turn around (hard on the back) or stand up (unsafe) to view your work. Belly blades are usually adjusted by hydraulics that raise, lower, or change the angle, tilt,

**This rear-mounted blade is handy for leveling gravel.**

**A plow mounted on a four-wheel-drive garden tractor is handy for moving light snow.**

**An old-time tractor with a belly blade.**

and pitch of the blade and can shift the entire blade sideways. If you are a serious grader, you can even add an optional laser system for leveling. It is important to check on the compatibility of a belly blade with your tractor and loader. Most belly blades will not work with cab model tractors.

Skid shoes are an optional bolt-on attachment that hold the blade off the ground a few inches to allow removal of snow or manure without digging into the road or pen surface. Another feature is that the cutting edge of many blades is either reversible and/or replaceable by removing bolts. Some blades have optional end plates that convert the blade into a box blade, which greatly assists in moving or leveling material. Or you can buy a separate box blade, which is a combination ripper and blade with endplates. The box blade has adjustable teeth (scarifier shanks) at the front of the box that dig into the soil up to four inches and a blade, or tailgate, at the rear that can be adjusted to float or grade. A box blade is most useful for landscaping, road maintenance, loafing shed cleanup, or ditch cleaning.

## Carts and Wagons

You can use the bucket of your loader to carry small loads of hay and bedding, and you can use your flatbed trailer for large loads. If you need something in between, consider a wagon or cart. A cart is basically a small trailer with sides, while a wagon is a long four-wheeled cart. Most models have a pin-type hitch. There are many sizes and styles to choose from, and this is one of the most useful implements you can have for an APV.

Two-wheel carts are the simplest and least expensive. Wagons usually have steerable front wheels, which make them easier to maneuver when backing. The load capacity of a cart or trailer will depend mainly on the size of its running gear (axles, bearings, tires, and suspension).

Carts are made of either plastic or steel. Steel carts are typically squarer in shape and have a flat bottom and a tailgate. Molded poly trailers are generally more rounded in shape, like a wheelbarrow, and most do not have a tailgate. A tailgate that opens can make it easier to load items and materials and to unload them without dumping. Some designs rely on the closed tailgate to hold the sides together or allow the sides to spread apart if the trailer is loaded with heavy material like dirt or gravel and the tailgate is open or not in place. Other models have reinforced sides to prevent spreading whether or not the tailgate is in place.

Most two-wheel carts have dumping capabilities, which means the bed can be tipped backwards to unload the contents. The bed usually dumps by means of a lever that unlocks the front of the bed from the tongue and then lifts the bed to pivot it backwards on the axle. Most carts dump manually, which means that you have to lift the front of the bed to tilt it back. A few heavy-duty trailers have a hydraulic dump powered by a small onboard electric

**A flatbed farm wagon has myriad uses including hauling hay, fencing supplies, and giving hay rides!**

motor that plugs into the tow vehicle to run the hydraulic pump.

Some cart models have a handle and folding tongue so the cart can be pushed or pulled by hand when not hooked to the vehicle.

If you don't have a tow vehicle, there are gas-powered walk-behind carts, wagons, and flatbeds designed to haul up to 800 pounds. An electric dump option is available on some models.

### HANDCARTS

Every horse farm needs one or two handcarts, depending on the particular use. Two-wheeled carts are more stable than wheelbarrows when it comes to moving heaping loads of manure or bedding. Other useful ones include:

- A lightweight garden cart for gathering and feeding hay dregs
- A medium-sized, lightweight but strong, aluminum-frame manure cart
- A heavyweight cart with tires for moving gravel and heavy loads
- Large-capacity wooden cart with bicycle wheels for moving hay and bedding

# Flatbed Trailers

A flatbed trailer is basically a long platform on wheels. The bed is typically comprised of 1 ½-inch boards or sheets of steel with some type of pattern for traction, typically diamond plate. Flatbeds are available in both straight-pull and gooseneck configurations (see chapter 9).

Common utility flatbeds have one or two (tandem) axles. The size of the axles determines how much weight the trailer can carry.

A flatbed trailer allows you to haul equipment and materials that are too large or heavy for your pickup or are easier to load onto a trailer. Flatbeds can be used around your acreage, for example, to deliver fencing materials where they are needed or to collect trees and brush that have been cleared. A flatbed can also be used to haul your tractor or other equipment to town for repairs, or to bring a large quantity of hay, bedding, or building or fencing supplies back to the barn.

Some trailers have ramps at the back for loading equipment. When traveling, the ramps either flip up and lock like a tailgate, or slide into channels under the bed. Tip trailers pivot at the center so that

**This gooseneck flatbed trailer can haul a year's supply of hay for a small horse farm and can transport equipment as well.**

the back of the trailer touches the ground to allow equipment to roll onto the bed.

The primary considerations when choosing a flatbed are weight capacity and size.

## FLATBED CAPACITY

One of the biggest mistakes people make is overloading their flatbed. Overly heavy loads can damage suspension, brakes, and tires, so make sure the trailer you buy is up for the tasks you have in mind. Flatbeds range in capacity from less than 1000 pounds to more than 50 tons, so there is bound to be one to fit your requirements.

Imagine the heaviest load you will need to transport, such as your 7000-pound tractor or five tons of hay, and look for a trailer that will carry that weight plus 1000 pounds more. The 1000-pound buffer gives you a margin of safety and results in longer trailer life.

To find out how much actual weight a flatbed trailer will handle, subtract the weight of the trailer from the trailer's gross vehicle weight rating (GVWR). For example, if the trailer has a 10,000-pound GVWR and the trailer weighs 2000 pounds empty, you can legally haul 8000 pounds of cargo.

Tandem axles, one in front of the other, can double the trailer capacity. Some people prefer tandem trailers because they follow the tow vehicle down the highway more smoothly and are less inclined to fishtail. The extra pair of tires also gives a smoother ride and safer handling in the event of a blowout. And if you do have a flat on a tandem trailer, you can pull or back the good tire onto a block or ramp to raise the flat off the ground for changing without a jack.

A single-axle trailer is less expensive to buy and to maintain because there are fewer tires, bearings, and brakes. A single-axle flatbed will tow just fine if you pay attention to how your load is distributed.

## FLATBED SIZE

Flatbeds range in size from 4-foot by 6-foot ATV trailers to 8½-foot by 36-foot equipment trailers. The physical size of the trailer is especially important if you are going to haul equipment. Make sure the deck of the trailer is wide enough so the tires on the equipment you are hauling don't hang over the

If a trailer has a GVWR of 7000 and the trailer weighs 1500 pounds, you can haul 5500 pounds of cargo or one hundred 55-pound bales of hay.

edge, and it is long enough to hold any implements that are attached.

Trailers come in both low- and high-profile models. Low-profile models make it easier to load and unload cargo, like bales of hay, because you don't have to lift so high. But the fenders on a low-profile trailer can make the center of the bed too narrow for some equipment and can affect how other cargo is loaded as well. In addition, the lower the trailer, the more likely it is to drag on either end when going over uneven terrain or dips in the road.

A high-profile trailer doesn't have fenders because the deck is above the tires. This makes for a wider deck that can accommodate a larger load. But the height of the bed can make it difficult to load both cargo and equipment.

## LOADING A FLATBED

There are three ways to get equipment onto a flatbed: backing the trailer up to a loading dock or earth bank, using ramps, or having the trailer tilt back so you can drive the equipment onto the trailer.

Loading off a dock or bank is convenient because you don't have to deal with ramps, but you are restricted to loading and unloading equipment in the same spot all the time. Ramps enable you to load and unload anywhere. Some ramps are the full width of the bed and flip up and latch in a vertical position like a tailgate for travel. Wide ramps accommodate vehicles of different widths and are convenient for carrying cargo on and off the trailer on foot.

Other trailers use double ramps, one on each side. They flip up and lock in place like a wide ramp or slide into channels under the bed when not in use. Some double ramps are adjustable for width while others are not. Make sure the ramp system will fit your equipment.

When you are shopping for a trailer, set up the ramps and put them away to see how easy they are to use. Take the trailer for a test ride to hear if the ramps rattle too much for your comfort level.

A tilt-bed trailer pivots over the middle so the back end of the trailer comes to the ground. This

A small tandem trailer is invaluable for hauling equipment and supplies and taking loads of trash to the landfill.

A tilt-bed trailer is easier to load than a fixed bed, especially for drive-on equipment.

## Trailer Axle Capacities

If you can't find the GVWR of a flatbed, which is often the case with used trailers, you can easily determine the axle capacity and thus the trailer capacity by looking at the wheels: Check the number of lugs or bolts that hold the wheels on the hub.

Generally speaking, an axle with five-bolt hubs is rated for 3500 pounds and uses 15-inch wheels; an axle with six-bolt hubs is rated for 5200 pounds and uses 15-inch or 16-inch wheels; an axle with eight-bolt hubs is typically rated for 6000 or 7000 pounds and uses 16-inch wheels.

The tires on a 3500-pound axle and a 5200-pound axle are almost always rated within the limits of the axle. When you look at 8-bolt wheels, check the ratings molded on the tire sidewalls to make sure they are high enough: 6000-pound axles require tires in the "E" range; 7000-pound axles require "H" range tires.

Warning: Don't substitute automobile tires for trailer tires, because trailer tires have thicker sidewalls that help control sway better.

enables equipment to drive onto the trailer and the trailer tilts back level when the equipment reaches the center of the bed.

### FLATBED OPTIONS

Side rub rails are smooth steel rails that run along the outside of the trailer to protect the trailer from snagging on objects you might run up against. They also help protect the object, a tree for instance, from damage.

Stake pockets are short sections of rectangular tubing attached at intervals along the sides of the trailer into which you can insert stakes or other dimensions of boards to stabilize the cargo. Plywood or other material attached to the stakes can form removable sides for the trailer. Stake pockets also provide a place to attach tie-down ropes, straps, and chains. A trailer without stake pockets should have holes, loops, or some other means of attaching tie-downs.

Look for rear lights that are recessed or protected from damage. If they stick out to the sides with no protection, you can count on replacing them often.

Small to medium trailers have a top-wind jack attached to the tongue for raising and lowering the hitch when hooking up. The jack typically swings up and fastens out of the way for traveling. Heavier trailers often have two jacks, one at each front corner of the bed.

Having brakes on all trailer axles can greatly reduce wear and tear on the tow vehicle brakes and increase stopping capability. As with a horse trailer, a breakaway trailer brake system on a flatbed automatically locks the trailer's brakes to stop the trailer if it becomes unhooked from the tow vehicle. (See chapter 9 for more on trailer brakes.)

# Auger

An auger, or digger, is positioned vertically and powered by the tractor to drill holes in the ground for setting fence posts, planting trees, and putting in poles for pole barn construction. The diameter of the auger and its resulting holes ranges from 6 to 48 inches. The depth of the hole is usually 42 inches, but some augers have optional extensions up to 14 inches long that can be added to create deeper holes.

Augers are either single or double flight, which means one or two sets of flanges spiraling up the shaft. If you are digging in sandy soil, double flight is okay, but if you are digging in rocky soil, single flight might be more appropriate so the rocks can fit in between the flanges and work their way up. Sometimes the term "flighting" also refers to the treatment of the cutting surface at the bottom of the auger. Double flighting basically means there are twice as many teeth as single. The teeth on the lower part of the spiral can be hardened steel wisdom or gage teeth or carbide teeth for rocky or frozen ground. The tip of the auger can be the standard, long-lasting spiral or a more aggressive fish tail point for tough digging.

Diggers usually offer shear bolt driveline or radial pin-clutch driveline protection, and often a hydraulic relief valve, all of which are designed to stop the unit when it hits an immovable object to prevent costly damage. Some diggers have handy stands that allow you to store them standing up. This not only makes hook up easier but protects the digger from deterioration by ground moisture.

A digger can attach to the tractor in several ways, most commonly to the three-point hitch at the rear of the tractor. This is less than ideal, however, because the weight of the auger and the three-point hitch may not provide enough downward pressure, which can make drilling in hard ground difficult. The temptation is for a person to lean on top of the auger to help it drill, but this poses an obvious safety risk and should never be attempted. Optional hydraulic down-pressure kits that work off the tractor's existing hydraulics can be installed to provide an additional 500 pounds of downward pressure for work in tough soil. A better option is an auger that attaches either to the bucket or, with the bucket removed, to the loader mounts, resulting in greater downward pressure from the tractor's hydraulic system.

Used diggers in good shape are hard to find because either the augers are worn or the gearboxes are shot. You can't rent the type of posthole digger that is used with a tractor, although two-man, gas-powered ones are generally available at equipment rental places. If you were just beginning to put in your facilities and had lots of fence posts in your future, it would pay to invest in a posthole digger for your tractor. Otherwise, it would be more economically sound to hire someone to dig the holes for you at a per hole rate. Or you can dig the holes by hand with a manual posthole digger. You'll need one of these anyway to clean loose dirt out of holes drilled by an auger.

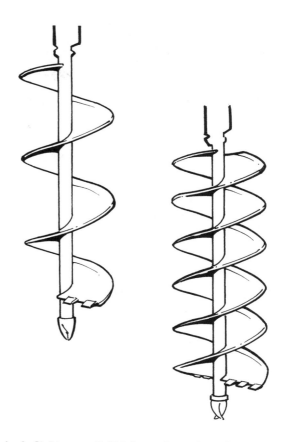

A single flight auger (left) is better for rocky soil while a double flight auger (right) is more suitable for sandy soil.

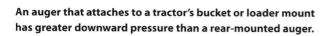

An auger that attaches to a tractor's bucket or loader mount has greater downward pressure than a rear-mounted auger.

# Making Hay

If you have sufficient acreage to grow pasture for grazing or hay, either for your own animals or to sell, you may want to invest in the following equipment.

### SPRAYER

A sprayer is useful for applying herbicide and fertilizer. As with spreaders, an electric sprayer either mounts on the back of a vehicle or is towed on wheels. Sprayer tank capacities range from 5 to 45 gallons. Attaching a boom to the sprayer enables you to spray a swath up to 50 feet wide, depending on the boom. This can make quick work of fertilizing a pasture or spraying weeds along a ditch. A wand or handgun with an adjustable nozzle is good for spot spraying smaller areas. A sprayer can be purchased to fit any size three-point hitch, depending on desired capacity and mode of operation.

### SEED DRILL

This combination implement aerates the soil, meters the seed, plants the seed into the earth, compacts the soil and, with an optional roller, can put the seed to bed! A seed drill is a more definitive way of planting hay fields and pastures, or reseeding or renovating acreage, than a broadcast seeder. It is especially beneficial in areas where wind or water erosion occurs.

### BROADCAST SEEDER/SPREADER

A multiuse unit, a broadcast spreader usually has a large steel or plastic hopper mounted above the spreading mechanism. Spreaders can be powered by the tractor, a separate electric motor, or its own wheels (ground driven). The hopper can range in size from 8 to 16 cubic feet for a 600–1000 pound capacity. It can be filled with seed, powdered or granule fertilizer, crystalline herbicides, and pesticides. It is handy for spreading sand or salt on icy driveways, too.

The spreading mechanism is PTO-powered, and the seeding discs have adjustments for various patterns and density of spread so you can choose the appropriate spread for the material you are using. Spreading width will vary from 15 to 50 feet depending on the product used and the wind. Broadcast spreaders distribute material in a wide pattern (up to 12 feet). Drop spreaders distribute material between the wheels (preventing overthrow), which makes them nice for controlled application of herbicides near shrubs, for example.

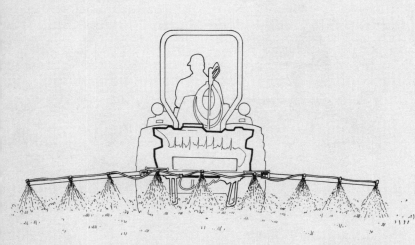

**Rear-mounted sprayer boom**

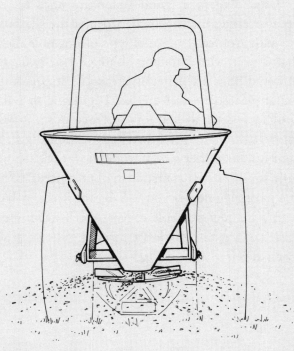

**Broadcast seeder/spreader**

**HAY BALER**

A baler is used to collect dried, precut grasses, alfalfa, and straw and form the loose piles into bales for convenient storage. A pickup trough collects the piles and moves them to an auger that positions them to be pressed into shape by a piston. When the pressed grass reaches a pre-set length, a mechanism signals the string to be tied and cut. The finished bale is ejected onto the field while a new one is started.

Purchasing your own hay baler, along with a mower, rake, field wagon, and other haymaking equipment, could run you $20,000, not including the cost of fuel, baling twine, and your time. Unless you are putting up hay on at least 40 acres, and enjoy doing it, it might make more sense to hire a custom baler to cut and bale your hay. You can hire someone on a cost per bale basis (perhaps $1.50 a bale) or you could make an arrangement to sharecrop. In that case, you supply the field of hay, and the custom farmer supplies the equipment, supplies, and labor, with the two of you splitting the crop (percentage varies widely).

Balers are intricate machines with many things that can go wrong and a deserved reputation for quirky behavior. If any piece of equipment is going to break down when you need it most, it will be the baler. My advice is that unless you bale a considerable amount of your own hay, or love fixing machines, stay away from balers.

# Disc

A disc turns over soil with rows of evenly spaced circular discs and is handy for working the soil in an arena or for aerating a very compacted pasture. Disc blades can have smooth edges or, for more aggressiveness, notched edges. For cropland, soil is first tilled, then disked, then pulverized.

A tiller is a rotary soil digger/mixer that works to eight inches deep and is more appropriate for cropland and gardens than pastures.

Pull-type discs, attached to the tractor's drawbar by a pin, are common and inexpensive, but they are difficult to transport as the arms cannot be raised and lowered like hydraulic or three-point discs. Some pull-type discs do have tires and a manual crank handle or hydraulic cylinder that allows you to raise or lower the disc as needed for road transport or cutting, but to be moved, most pull-type discs either have to be loaded onto a trailer or dragged behind the tractor, disking everything along the way including driveways, road surfaces, and grassy areas! In addition, if you use a pull-type disc to work an arena, you will not be able to back into the corners and will end up with an oval area of worked soil.

Hydraulic discs can be raised off the ground to prevent damage to areas you don't want worked and to position the disc in tight spots. You can raise the disc, back it into the corners or up to the fence line of your arena, field, or pasture, and work the entire area. Smaller hydraulic discs (six-foot) operate off the tractor's three-point hitch and are raised when moving from point A to point B. Larger, heavier

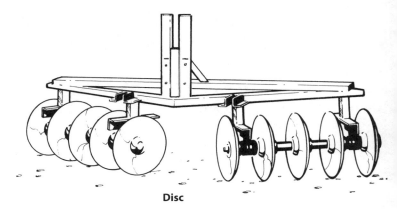

**Disc**

discs have tires that are lowered to carry the weight of the disc for positioning the implement and for transport.

Both a hydraulic disc and a pull-type disc do an equally good job, but an average six-foot three-point disc costs about three times as much as a pull-type disc. Avoid old discs with box bearings because box bearings are very difficult to find today. It is best to stick with sealed bearings.

## Backhoe

A backhoe with a narrow bucket is most often used on a tractor or skid steer for digging trenches and ditches for utility lines. A wider bucket is used for larger excavations like cleaning ponds or digging graves. Buckets are available in widths from 9 to 36 inches and a typical backhoe can dig eight feet deep. If you don't have much trench or ditch work, it would make more economic sense to hire someone to dig for you or to rent a self-contained backhoe from your equipment rental shop.

**A backhoe is ideal for digging trenches.**

## Snowblower

A snowblower displaces snow farther than a blade or plow. A blade and a bucket work fine in some instances, but you can soon run out of places to pile the snow. With a snowblower, you can blow snow up to 75 feet, giving you a much larger area in which to dispose of heavy snows. Unlike blades, snowblowers don't leave berms blocking doors and gates that you have to shovel by hand.

The terms "snowblower" and "snow thrower" are sometimes used interchangeably, but there is a difference between the machines. While both have a horizontal auger that gathers snow to the middle, a snow thrower is a single-stage machine in which the auger spins very fast to bring snow to the middle and throw it up the discharge chute. A snowblower is a two-stage machine in which a slower auger brings snow to the middle where a high-speed impeller blows the snow up the chute. Since the auger of a snowblower spins more slowly than that of a snow thrower, it is less likely to be damaged by rocks or other objects. A two-stage machine can blow snow farther and is more effective in all types and depths of snow than is a single-stage machine.

A snowblower can be fitted to a tractor, skid steer, and many APVs. It can be powered by the vehicle's PTO or hydraulic system or its own gas engine. Adjustable skid shoes enable you to control the cutting depth, which is especially useful on gravel driveways. A reversible scraper blade, or a wear bar, on the cutting edge will greatly extend the life of the edge. Drift blades are vertical extensions attached to the housing that act like knives to leave a clean edge when cutting through drifts and snow banks.

One of the best features of a snowblower is that you can control where the snow goes. The discharge spout and deflector can be adjusted to direct the plume of snow discharge. You have to do this manually on some models, while on others you can move the spout from the seat hydraulically or by means of a cable. *Note:* It is a good safety precaution to use tire chains on any wheeled vehicle that is operating a snowblower.

# Miscellaneous Implements

Depending on your acreage and the type of land you have, these implements may be worth renting or even purchasing to accomplish major tasks such as clearing trees or rehabilitating pastures.

### AERATOR

An aerator is essentially a rotary drum with spikes to poke holes in the ground as it rolls forward behind the tractor. If you have compacted soil that you don't want to plow, it can be aerated to improve uptake of water, nutrients, and air, which improves root structure and plant growth. An aerator can be used routinely for pasture maintenance or prior to reseeding bare patches.

### BROOM

A rotary broom, or sweeper, is like a huge version of the broom on a carpet sweeper. The cylindrical broom is typically 24 inches in diameter and 60 inches wide and has both synthetic and steel bristles. A rotary broom can clear dirt, snow, and debris off of hard surfaces, as well as de-thatch and sweep lawns clear of leaves and cut grass. It might be a useful tool if you have large areas of concrete like a parking lot or walkways that you'd like to keep clean.

Using this type of broom in a barn, however, will kick up a lot of dust. A lawn or shop vacuum works much better for picking up debris and dust in the barn aisle.

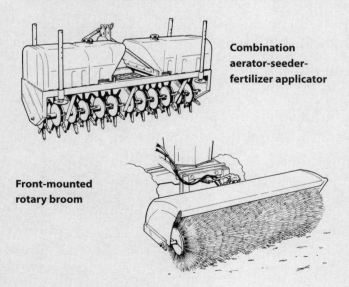

**Combination aerator-seeder-fertilizer applicator**

**Front-mounted rotary broom**

A rotary broom either has its own gas or electric motor or is powered by the tractor's PTO or hydraulic system. Some brooms can be angled, like a blade, for better control of material movement. For tough jobs, like cleaning packed mud off concrete, a stiffer broom with all-wire bristles is available. A rotary broom of sufficient size can throw an 8-inch layer of snow up to 15 feet and can clean packed snow down to hard surface.

A lawn sweeper is towed behind a vehicle. It is not a vacuum, but has a series of revolving brushes that pick up lawn cuttings, leaves, twigs, and other debris and send them into a bag or bin. The brushes are powered by the wheels, so look for a model that is designed for maximum traction.

### WOOD CHIPPER

If you have a lot of trees or brush on your property that you plan on clearing or trimming, a chipper might be a good investment to shred the branches and limbs. This process will quickly reduce the volume of the debris, making it easier to dispose of or use as mulch.

Chippers are usually gas powered, but there are electric and PTO models available. Most chippers can handle branches up to 4½ inches thick and can pulverize weeds, leaves, and garden refuse into mulch and compost. Some chippers are designed for mounting on a three-point hitch, some are coupled with a lawn vacuum, and some have a wheeled frame for towing or moving by hand like a wheelbarrow. A top discharge chute is handy for directing chips into a truck for hauling.

### STUMP GRINDER

A stump grinder would be useful if you need to clear out a lot of large trees or if your property contains many stumps. For a smaller job, it might make more sense to rent one. A stump grinder has a high-speed wheel with carbide teeth that grind the stump away in small increments as the machine is lowered onto the stump. Nothing is left but a pile of chips. Stump grinders are available for small to large tractors and are powered by their own engines or the tractor's PTO and hydraulics.

# 5 All-Purpose Vehicles

Depending on your vehicle needs, you might want to consider an all-purpose vehicle (APV) instead of, or in addition to, a tractor. APVs include both all-terrain vehicles (ATVs) and utility task vehicles (UTVs). Most APVs are four-wheel drive and some feature a cargo box like a small pickup bed that has room for hay and grain. Some models can haul up to 1000 pounds. APVs take less space to store and are usually less expensive to purchase and operate than a tractor.

## What an APV Can Do

An APV can do just about everything a tractor can do, but on a smaller scale. Since these vehicles are smaller and lighter than a tractor, they do some jobs — like moving snow or dirt — more slowly and less efficiently. And they simply do not have the power to do heavy hauling, or fieldwork like baling hay, or pulling large cultivating equipment. An APV's smaller size is an advantage for other chores, however, and they can be used for many jobs around an acreage where a tractor or a pickup would be too large or inappropriate for the task, such as traversing hilly terrain to mend fence, carrying feed to horses on pasture, or cleaning small pens.

## Comparing ATVs and UTVs

The ATV is basically a four-wheeled motorcycle, while the UTV looks like a small pickup truck and is an offshoot of both the ATV and of the golf car. The UTV is also called a cargo all-terrain vehicle or a turf truck. Although an ATV and a UTV can do many of the same jobs, there are important differences between them.

- An ATV has handlebars; a UTV has a steering wheel.
- An ATV has a seat you straddle; a UTV has a bench or bucket seat you sit on.
- An ATV is designed for one rider and very little cargo; a UTV is designed for two or more riders and a significant amount of cargo.
- An ATV is generally smaller and quicker than a UTV.
- An ATV costs between $3000 and $8000 depending on size and options. A UTV can run from $5000 to more than the cost of a full-sized pickup truck.

Do you want to straddle an ATV or sit on a UTV seat? When traversing rugged terrain you might feel more secure on an ATV with a leg on either side and holding on to handlebars rather than sitting on the bench or bucket seat of a UTV. An ATV is also quicker and more maneuverable for agility chores like herding cattle or other animals.

UTVs are generally heavier and less nimble than ATVs, but they are easier for less flexible people to get on and off. They can carry more cargo on the vehicle itself, but you can always add a trailer behind an ATV and pull just as much cargo as most UTVs can carry. The choices and combinations are many and varied.

**A UTV makes a handy vehicle for feeding horses and some think the UTV *is* the feeder!**

**An ATV rides like a motorcycle — it is quick and maneuverable but has less cargo capacity than a UTV.**

This UTV has a roll bar structure similar to the ROPS on a tractor.

## Golf Cars — Flatland APVs

Golf cars, or carts, were designed as safe, easy-to-use rental machines for rolling over the smooth grassy areas of a golf course. Many farmers and ranchers have found them useful as low-maintenance chore vehicles — as long as they stay on smooth ground. Golf cars are built low to the ground to make it easy to get in and out of them and to minimize the chance of the vehicle overturning. This low clearance means the vehicle can easily get high-centered on ridges, hummocks, and rocks around a horse acreage. You can raise the clearance by installing larger tires or a lift kit, or both, but expenses for modifications can add up and may put you in the price range of an APV that would better fit your needs.

# APV Features

Whether you are leaning toward an ATV or a UTV, there are many features to consider: cost, comfort, style, safety, load capacity, agility, toughness, service, warranty, options, and attachments. The choices can be overwhelming. Rather than getting bogged down in the details, start by looking at several basic features: size, payload capacity, engine type, and drive-system type.

## SIZE

Generally, the longer the wheelbase, the wider the track, and the lower the center of gravity, the more stable a vehicle is. The narrower and higher a vehicle is, the less suitable it is for hills and rough terrain. A shorter wheelbase results in a smaller turning radius and more maneuverability, while a longer wheelbase provides a more comfortable ride. Track width is the distance between the centers of the tires along the axle. All other factors being equal, a vehicle with greater track width is less likely to tip over.

The curb weight of APVs varies from less than 500 pounds to more than 1700 pounds. A heavier vehicle is more stable and less likely to tip over when towing a trailer or equipment and when using attachments like a loader or blade. More weight also generally means a more comfortable ride.

Length, width, and height are important considerations when selecting an APV. The overall length of APVs varies from 60 inches to 148 inches. This helps

Excellent for use in small spaces and quiet so as not to scare horses, a golf car can be very handy.

An ATV with a sprayer tank makes the job of spot-spraying weeds along driveways and fences downright fun!

you determine how much room you'll need to store the vehicle. Width, along with length, affects parking space, but more importantly, it dictates what openings will accommodate the vehicle. Overall width varies from 34 inches to 61 inches, so some models will drive through a standard doorway, while wider models might not even fit through a five-foot gate. If you have many gates to go through, you would be better off choosing an APV that fits the gates rather than widening all the openings.

Width is also a factor if you'll need to transport the vehicle to use it in more than one location. All ATVs and many UTVs will fit into the back of a full-size pickup truck. Even so, you'll need to figure out how you load and unload it. You might be able to find a suitable earth bank you can back your truck up to use, or you can buy a special ramp or build a loading dock. Some UTVs are almost as large as a full-size pickup, so you would need a flatbed truck or a trailer to haul one of those.

The height of APVs varies from 44 inches to 81 inches to the top of the roof or the rollover protective structure. Make sure to measure the heights of any openings you plan to drive the vehicle through for work and for storage. More than one proud owner has brought a new vehicle home only to find it wouldn't fit into the storage building.

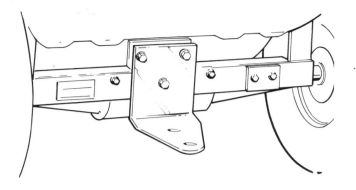

**Most APVs come with a simple pin hitch for towing a trailer or cart.**

## TOWING AND LOAD CAPACITY

The amount of weight an APV can tow ranges from 400 to 4000 pounds. Towing capacity is figured for pulling a wheeled vehicle, like a trailer or cart, on a dry, level surface. Towing capacity drops quickly when going uphill or operating on slippery footing. APVs are very useful for pulling trailers of various types, from a flat-bed for feeding hay or hauling fence posts and wire to the pasture, to a dump-bed trailer for moving dirt, gravel, or loose bedding.

Trailer sway is not as big an issue with APVs as it is with pickups, because APVs tow at relatively slow speeds, but tongue weight can affect your control in other ways. Just as with a tractor or truck, the correct tongue weight optimizes traction on the rear wheels. The total weight of the load plus the trailer should not exceed the rated towing capacity of the vehicle. (See chapters 8, 9, and 11 for more information on hitches and towing safety.) When towing heavy loads with an APV and going downhill, it is safest if the trailer or equipment that is being towed has its own brakes.

Depending on what you plan to haul on the vehicle itself, load capacity can be an important factor in choosing an APV. In order to compare load capacity among vehicles, you have to know exactly what the listed load capacities include. With ATVs, the load capacity almost always is stated as cargo capacity and refers to how much cargo you can put on the front and on the rear cargo racks (these numbers

**A lightweight aluminum ramp made loading this ATV a snap.**

are often listed separately). The largest ATVs have a cargo capacity of 100 pounds on the front rack and 600 pounds on the rear rack.

With UTVs, the terms "load capacity," "cargo capacity," and "payload" are sometimes confused. Payload, or total payload, refers to the total weight the vehicle can carry and typically includes the weight of the vehicle, two 200-pound passengers, and total cargo weight. Cargo payload, or bed payload, is the amount of weight you can haul in the box or bed. Different manufacturers define load capacity differently, so ask the dealer to explain it for the models you are considering.

Cargo capacity for UTVs ranges from 400 to 2000 pounds. ATVs can't carry as much, but the largest ATVs are rated to pull a trailer weighing 1050 pounds. Consider all the components of a vehicle when determining what you need for hauling cargo. To help decide how much payload capacity you'll need, figure out the weight of the heaviest materials you might transport with the vehicle. A standard square bale of hay, for example, typically weighs between 50 and 65 pounds. Most UTV cargo boxes are large enough to hold two bales across and up to three high, a total of 6 bales or 380 pounds — well within the range of most UTVs and ATV trailers.

Just because a vehicle has a large cargo box, however, doesn't mean that you can fill it with gravel and expect the suspension and engine to handle it. If the

cargo box holds 11 cubic feet, a full box of gravel (at 90–145 pounds per cubic foot) adds 1600 pounds to the vehicle — well over the limit of all but the largest UTVs. Heavy work like this is usually better suited to a truck, tractor, or loader.

## Cargo Box

The cargo bed or box is the main reason that many people buy a UTV. A bed is just a flat surface, while a box has formed sides of varying heights. The size of the box, how it is configured, and how easy it is to use will determine how useful the vehicle is for you and how well you'll like it. The dimensions of the cargo area are especially important if you plan to haul bales of hay and a large volume of other materials. Cargo box length varies from 42 inches to 61 inches; width from 48 inches to 58 inches; depth or height of the sides from 9 inches to 13 inches. Capacity varies from 11 cubic feet to 18 cubic feet.

The height of the cargo area from the ground determines how difficult it is to lift cargo into it. This height can vary from 25 inches to 36 inches, which is quite a spread. A lower bed makes it easier to load and unload manure and hay bales and feed, and helps keep the center of gravity low, meaning less chance of tipping over on a hill or rough terrain. Practice lifting things into the vehicle at the dealer to see how the bed height works for you.

Other options to look for include hooks or loops on the bed that allow you to attach ropes or tie downs to keep your cargo in place, and holes for stakes or side extensions to help contain bulky loads like straw or bedding. You might find it easier to use a tailgate that folds down rather than one you have to lift out and then lift back in place, although a removable gate is nice if you seldom need to close it. It depends on the work you need to do.

A powered lift, or dump, bed or cargo box is useful for unloading material such as manure, bedding, mulch, sand, and gravel so you don't have to shovel it out or unload it by hand. A power lift operates either hydraulically or electrically. Some vehicles only allow you to operate the lift from the cab. With some

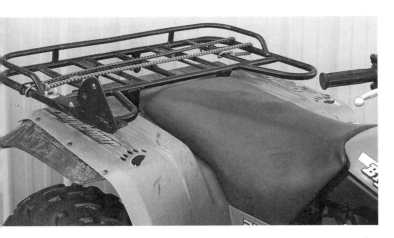

**Although limited in size, cargo racks on an ATV can carry tools and supplies and even a bale of hay or bedding.**

**A dump bed makes it easier to unload manure and bedding.**

electrically powered beds you can stand beside the machine and use a remote control to raise the bed, with the motor off and brakes locked. These setups allow you to stand where you can get a better view of where you are dumping.

## Seating

An ATV traditionally seats one person, although a few newer models seat two. Some UTVs can seat up to six. Having room for two or more people makes the vehicle useful for hauling tools and workers out to fix fence, for example, or for taking visitors to the lower 40 for a picnic.

Adjustable seat position and the height and design of seat backs can be important, especially if you are getting the vehicle to make your life more comfortable. Compare those features during your test-drives. Seats that fold down or are easily removed make a more versatile cargo space.

Seat belts are not appropriate for an ATV, and guidelines for their use with a UTV are the same as with a tractor. (See rollover protective structure on page 28 and also in chapters 2 and 11.)

## ENGINE

Unlike new tractors, which are only available with diesel engines, some APVs are available with gasoline, diesel, or electric engines. Some gas engines can be fitted with a conversion kit to run on propane or natural gas.

The engines in some APVs are sealed so well that you can drive through more than 16 inches of water. This could be a valuable feature if there is a creek on your property that you need to ford or if you live on a flood plane.

Gasoline engines are available in ATVs, UTVs, and golf cars. Three reasons for choosing a gasoline engine are convenience, economy, and noise level. If you are already storing gasoline on your property for lawn mowers and other equipment, another gas engine will fit right into the fuel plan. A gas engine is less expensive to buy than a diesel engine. It is also easier and cheaper to repair. It will produce more horsepower per pound than its diesel counterpart. Gas engines typically have a lower noise level than diesel engines and the exhaust is not as offensive to as many people as the exhaust of a diesel.

Diesel engines are available in some UTV models. Three reasons for choosing a diesel engine are power, reliability, and longevity. (See chapter 2 for more on diesel engines.)

Three reasons for choosing an electric engine (some people use the term electric motor, but there's no law) are convenience, economy, and environmental impact. Electric engines are rare on ATVs, but are an option on several types of UTVs and most golf cars. New electric vehicles are no wimps. Golf car electric engines are commonly rated at 2–3 hp and they can produce 10–12 hp for a short time — enough to move a 5000-pound trailer and more than any similar gas or diesel APV.

Ask any golfer — electric vehicles are the easiest vehicles to use, as you just get on and go. There is no engine to worry about starting in cold weather, and you don't have to turn them off to save fuel because they don't idle. They run very quietly, making them a good choice if you have neighbors who don't share your active lifestyle. One thing to keep in mind is that some electric UTVs do not have rear suspension so they can ride pretty rough over anything but smooth ground.

Unlike most gas vehicles, many electric vehicles have neither transmission nor clutch. Speed is determined by how much electricity is sent to the engine via a foot pedal. Golf cars have mechanical brakes (no hydraulics) and only on the rear wheels.

Electric vehicles require less maintenance than gas or diesel ones, because there are no tune-ups, spark plugs, oil changes, filters, belts, carburetors, fuel pumps, or mufflers. You must check the water level in the batteries and keep the battery terminals clean of corrosion. But these savings are offset by the fact that all the batteries (usually three to six in number) have to be replaced every three to six years, at a significant cost. With proper maintenance, however, some batteries can last 10 years. As long as you have access to a standard 110-volt outlet, you will always have access to cheap fuel. Some vehicles can be even charged directly from solar, hydro, or wind systems. Depending on how low the batteries are, it could take up to 12 hours to fully recharge. If you plug it in overnight you'll be ready to go again in the morning.

An electric vehicle can run at top speed for one and a half to three hours before needing a recharge, depending on the type and condition of the batteries. Around a small acreage under normal use and slower speeds, that operating interval could provide more driving time than you would need in one day. If you need a vehicle to do steady hard work all day long, a gas or diesel model might be a better choice. If you run out of gas you can always hike back and get some, but if your batteries go dead you either have to tow the vehicle or haul the heavy batteries in to charge them and then haul them back to vehicle. Fortunately, most electric vehicles have a gauge that lets you know how much charge is left in the batteries. Another advantage is that with a simple converter or reducer, you can run electric power tools off the batteries.

## Cylinders and Power

A cylinder is a chamber in the engine block of a gas or diesel engine in which a piston slides up and down and displaces air and/or fuel to power the engine. APV engines have one, two, or three cylinders. More cylinders do not necessarily mean more power. A single-cylinder engine can have more power than a two-cylinder engine that has less displacement. APV engines vary from 9.5 to 56 hp, but are often categorized by displacement rather than horsepower.

A stroke is the movement of a piston in one direction. A cycle is comprised of two or four strokes. A cycle in a four-stroke engine occurs in two revolutions of the crankshaft:

1. **Stroke down:** takes fuel/air into the cylinder

2. **Stroke up:** compresses fuel in the cylinder

3. **Stroke down:** ignited fuel forces piston down, providing power

4. **Stroke up:** forces exhaust from the cylinder

The single cylinder four-stroke engine is considered by many to be the ideal power plant. It uses approximately 50 percent less fuel than a two-stroke engine, is highly reliable, and has a high resale value.

A two-stroke engine combines the operations of four strokes into two, an intake/compression stroke and a power/exhaust stroke, both of which occur in one revolution of the crankshaft. A two-stroke engine is cheaper and simpler, having one-third the moving parts of a four-stroke, which also helps make it very lightweight (ideal for chain saws and weed trimmers).

While four-stroke engines burn straight gasoline, two-stroke engines require a mixture of gas and special oil. The gas/oil mixture does not burn as cleanly as plain gasoline, so two-stroke engines tend to smoke, foul their ignition systems, and leave an oily residue on nearby surfaces and oil spots on the ground. Note: Gas with methanol can damage some two-stroke engines.

## Cooling

Combustion engines in APVs can be air-cooled or liquid-cooled. An air-cooled engine has fins on the cylinders that dissipate heat to the air. It also has shrouds that direct moving air over the fins and help to muffle engine noise. Be aware that the shrouds can collect gas and oil and dust, especially on a two-stroke engine, and might become a fire hazard.

A liquid-cooled engine has a pump that circulates coolant (typically water and alcohol) through channels in the engine and through a radiator to remove heat — just like the engines in trucks and autos. It has no bulky fins and shrouds, so takes up less space for an engine of the same horsepower. But at the same time it is heavier than an air-cooled engine because of additional parts and coolant.

Liquid-cooled engines cost more initially because they have more complex components. But they are more reliable, easier to maintain, and last longer. And they are quieter because the coolant absorbs sound. The big drawback of coolant is that it is a hazardous waste — it will kill plants that it contacts and animals that ingest it. Coolant must be dealt with responsibly if the system should leak or when the coolant needs to be drained for replacement or service.

Both air-cooled and liquid-cooled engines can supply engine heat for a UTV cab heater. An air-cooled engine pulls in heated air from the engine compartment and can bring in unpleasant and noxious fumes as well. With a liquid-cooled engine, heat for the cab is taken from water lines passing through the heater, which is like a small radiator, so the heat is fume-free.

## DRIVE SYSTEM

The size and configuration of the tires, transmission, and differential determine to a great extent the pulling power and top speed of a vehicle. Most APVs have a top ground speed of 25 mph, but some models can go faster than 50 mph. Slower models don't necessarily lack the power to go fast, but manufacturers install devices that limit how fast their vehicles can go to prevent personal injuries and vehicle

damage. For safety, the top speed of golf cars is limited to 18 mph, regardless of engine size.

## Transmission

An APV transmission should at least have gears for forward, neutral, and reverse. Many vehicles also have an auxiliary gearbox with high-low range, and some have a middle range as well. In addition to manual and automatic transmissions, you can choose an electric shift, which allows you to shift up or down with a button or switch, or a hydrostatic or variable hydro transmission (VHT) with infinitely variable speeds. Another option is a continuously variable transmission (CVT). Like a VHT, a CVT adjusts to maintain constant engine speed while delivering varying wheel speeds. This results in better fuel economy and smooth performance — and you can't feel it shift.

A shuttle lever is as handy on a UTV as it is on a tractor for quickly going from forward to reverse and back without using the transmission to shift gears. This is a very nice feature when you are using a loader and constantly going forward and backward to move snow, dirt, or manure.

Engine braking in a manual transmission allows drag from the engine to slow the vehicle when going downhill. This can save wear on the brakes and give you more control. Some automatic transmissions allow engine braking, while others do not. Engine-braking capability in electric engines, which do not have transmissions, is called "dynamic" braking, but not all electric engines have it.

## Differential

Differential lock is a standard feature on some APV models and optional on others. It operates manually by a hand lever or foot pedal, or electronically by a button or switch. Some types can be engaged while the wheels are moving; with others the wheels must be stopped. Some models have automatic differential lock that engages when needed and then unlocks for turning. (See chapter 2 for more information on differentials.)

## WHEELS

ATVs have four wheels, while UTVs can have four or six wheels. Six-wheelers have two front wheels and four rear wheels. You might come across an old three-wheel or five-wheel vehicle with a single front wheel for steering, but manufacture of these ceased by the early 1990s for safety reasons. Two-wheel-drive models have power going to only the rear wheels, which works well for light work in undemanding circumstances. If your acreage is relatively level and you don't get much snow, two-wheel drive will likely meet your needs. You'll get a lighter vehicle for less money. But if you do choose two-wheel drive, consider getting a limited slip or locking differential.

For use on rough, steep, muddy, or snowy ground, four- or six-wheel drive is in order. A few four-wheelers have full-time four-wheel drive. Many models let you switch between two-wheel drive and four-wheel drive with the push of a button. Some six-wheelers have the option of rear two-wheel drive, rear four-wheel drive, or six-wheel drive. On-demand systems have sensors that automatically engage the

This 6 x 6 UTV can carry cargo and passengers over varied terrain with ease.

front-drive axle when the rear wheels begin loosing traction.

Four- and six-wheel-drive vehicles generally have more powerful engines with a lower gear ratio to engage all four wheels. They can haul and tow heavier loads and withstand more abuse. Six-wheel vehicles have the best traction of all and, with two extra wheels to distribute the weight, they can handle the heaviest payloads. Because the weight is distributed over six tires, they leave shallower tracks. In soft earth or mud your footprint will likely be deeper than the tire tracks.

### Tires

ATVs and UTVs generally use tires from 22 inches to 25 inches high and you have a wide choice of tread designs. Standard size for golf cars is 18 inches, which is one reason the cars ride so low to the ground. As with tractor tires, APV tires are available in a variety of tread patterns.

High-flotation tires have extra strong sidewalls so they can run with very little air pressure. This makes them "soft" and spreads the load over an even greater surface area, making for minimal tracks. They cost slightly more than regular tires.

Run-flat tires have special sidewalls that are tough enough to let you drive on the tire without air for a limited time without damaging the tire or the wheel. This means you don't have to use any of your vehicle's limited cargo space to carry a spare tire.

## STEERING

With rare exceptions, front-wheel steering is found on all APVs. In the showroom and during the test, you can check out the feel and the position of the ATV handlebars or UTV steering wheels to see how they suit you. A steering wheel that tilts or telescopes, or both, enable a comfortable fit for different sizes and body styles of drivers who might use the vehicle. The diameter and the thickness of the steering wheel also can affect how comfortable and secure the wheel feels.

Turning diameter is the smallest size circle the

vehicle will turn. A smaller turning diameter makes the vehicle easier to maneuver in small spaces such as a barn aisle or a pen. It is sometimes expressed as turning radius (half the turning diameter). Turning diameters range from 12 feet to 26 feet. Generally, a smaller vehicle has a smaller turning diameter. If you are cleaning a 16-foot-wide pen, for example, a vehicle with a 12-foot turning diameter will be able to turn around in one motion, without backing up and going forward. Turning a vehicle with a turning diameter of more than 16 feet will take at least one back-and-forth maneuver, if not more.

### MATERIALS

The materials used for the body and for the frame of a vehicle determine how much the vehicle weighs, how long it lasts, and how easy it is to maintain. The body of a vehicle, including the cargo box, can be steel, aluminum, fiberglass, or plastic (commonly ABS composite: acylonitrile-butadiene-styrene). Steel is strong, but it is the heaviest and will rust. Aluminum, plastic, and fiberglass are lightweight and don't rust, so would be better choices than steel for use in humid environments and along saltwater coastlines.

Plastic and fiberglass are the least affected by the corrosive chemicals found in fertilizers and herbicides and will not dent like steel or aluminum, though they may crack. Plastic often is colored all the way through so is less likely to show scratches than is painted material.

The frames of most APVs are made of steel, but some manufacturers use aluminum to reduce the weight of the vehicle. As with the bodies, aluminum is less likely to corrode from chemical on fertilized grass, for example, or from salty coastal air.

A few models have an articulating frame that hinges in the middle to allow the vehicle to twist and flex when going over bumps and holes. With a rigid frame it is easy for one wheel to come off the ground on uneven ground. An articulated frame helps to keep all four wheels on the ground for better traction and stability.

# Options and Accessories

There are many other features worth considering that differ between models of ATVs and UTVs. Some can affect safety, while others could make the difference between a satisfactory purchase and buyer's remorse.

Most APV manufacturers offer small accessories like racks, bags, winches, grips and hand guards, windshields, vehicle covers, helmets, trailer hitches, and maintenance items, but they don't make or sell very many large attachments or implements.

### SAFETY OPTIONS

A rollover protection system is available for UTVs but not for ATVs and is similar in design and function to a tractor ROPS. A grab rail is a handhold on UTVs that provides additional security for passengers when traveling over rough terrain, especially if seat belts are not used.

A handheld spotlight or work light, or one that mounts on the sides or top, can help illuminate tasks after dark and increase visibility for night driving. For nighttime emergencies such as tending to injured stock or making repairs, your UTV can act as a portable light stand.

Full light kits that include stop/tail/turn signals and yellow flashers can increase safety in heavily populated areas or when driving in traffic. (Many state and local traffic regulations prohibit operating APVs and UTVs on public roads, so check before going off your property.)

A back-up beeper is a very good idea to warn people when the vehicle is backing up. Being backed into or over is a major cause of accidents with small vehicles.

A horn can be useful for alerting other people if you get in a bind, for moving livestock out of your way, or for signaling animals to come for feed.

Mirrors can help you see what is going on behind your vehicle without craning and straining, such as when you are towing a trailer or sprayer or mower.

An exhaust spark arrestor comes standard on some models. It can prevent the vehicle exhaust from starting fires in dry grass or brush and is an especially good idea in areas of high fire danger.

A stake kit allows you to put vertical stakes at intervals along the sides of the bed or box to stabilize cargo and keep it from sliding around or falling out during transport.

A cargo net is a quick way to secure a load from sliding around or blowing away. Throw it over a bunch of boxes or a load of brush, tie it down, and away you go.

A back screen or heavy metal grill between the cargo area and the seat prevents cargo from crashing into you, offers some protection from the sun, and provides a place to attach head rests.

Front and rear bumpers, which are standard on some models, can protect people and animals from injury should the vehicle run into them. Bumpers can also prevent damage to the vehicle in case of collision or when going through thick brush.

Bush or brush guards come up higher than the bumpers and often wrap around farther to offer even more protection.

A skid plate is a smooth metal plate that covers the belly of the vehicle for protection against damage from stumps, rocks, and other debris.

## BELLS AND WHISTLES

Most of the accessories listed here are only for UTVs, because they tend to be larger than ATVs and have more places to attach frills. By the way, don't expect the array of colors you're used to choosing from when buying a car or truck. Choice of ATV and UTV colors is very limited, partly because manufacturers want to maintain product identity.

A top or canopy to protect driver and passengers from sun and weather can be installed on UTVs and golf cars with or without a ROPS. A windshield is especially useful when traveling faster than 20 mph, because it keeps bugs, dust, rain, and snow off your face, and it is nice to have at any speed when the temperature is below freezing. It could improve your visibility in fog or light rain. Some windshields are fixed and some fold down out of the way. Floor mats can help prevent slipping on the floor in wet, snowy, or muddy conditions and make it easier to clean mud, snow, and hay from the floor of the cab. They also can make the cab quieter.

A molded or sprayed-on bed liner, or a fitted mat, can help keep cargo from sliding around and make cleanup easier. It will also increase the life of the bed by protecting it from dents and scratches. Liners are replaceable unless they are sprayed on, in which case you could have it resprayed.

A tool holder is a rack with hooks and slots and holes for organizing and securing cumbersome items like ladders, rakes, shovels, saws, and weed trimmers. It can protect both the tools and your cargo bed from being damaged.

It seems you can never have too many storage compartments. A glove box can hold gloves, glasses, notebook, and camera. Other storage boxes are useful for hammer, pliers, screwdrivers, wrenches, and other tools. For ATVs you can get hard cargo boxes, soft cargo bags, and an extra rack for the front of the vehicle. A bed enclosure turns a UTV into a small van and is handy for protecting cargo from weather and wind. Specially designed dog crates enable you to take your dogs along without having them running wild.

A propane conversion kit allows certain gasoline engines to run on more environmentally friendly fuels like propane or natural gas.

Mud flaps can help keep the vehicle, the driver, and towed equipment cleaner when driving through mud or rain. They also help keep flying stones under control.

If you have a very large ranch, or like to go where no man has gone before, you might find a global positioning system (GPS) option useful. By communicating with three of many orbiting satellites, a GPS can pinpoint your location on the earth with accuracy ranging from 10 meters to less than a meter, depending on the system. A GPS could be useful for mapping your property or laying out fence lines.

Cruise control is not only for highways; it also can be a useful option to maintain a steady speed when mowing, broadcast spraying or when spreading seed, sand, or salt.

A 12-volt power port (cigarette lighter) will enable you to run an air compressor, heater, and other 12-volt tools. An electric generator and/or a converter will let you operate standard 110-volt power tools like saws and drills.

Cab enclosures, either rigid or of canvas and clear plastic, offer protection from wind, rain, snow, and insects. A cab is especially nice to keep mosquitoes away when doing chores at dusk.

Along with a cab enclosure, a cab heater and/or air conditioner make you more comfortable when you are in the cab for extended periods, like when clearing snow or mowing. Add a sound system and you'll be praying for a blizzard or for more rain to make the grass grow so you can spend more time in the cab of your UTV.

For winter riding on an ATV, heated grips can keep your hands from getting cold. A rumble seat is a detachable rear seat on a UTV that can be useful for times when you need to haul a fence crew to the pasture or give visitors a tour of your place. A cup holder can free up both hands without you having to grip your beverage between your legs.

## IMPLEMENTS

There is an astonishing variety of implements made to fit APVs, many of which will fit lawn and garden tractors as well. But just because an implement will attach to a vehicle doesn't mean the vehicle can handle it. For example, a forklift or bucket, plus the load it carries, will add a lot of weight — maybe too much — to the front or rear suspension of an ATV. This not only limits the lifting capacity of the attachment but also could damage the vehicle and be unsafe. Be reasonable about your expectations.

The majority of APVs have no means of supplying power to attachments, so those implements that require power have their own engine, usually gasoline powered. Only the largest UTVs are specifically designed to use and to power heavy-duty attachments like augers, loaders, snowblowers, and mowers, an ability that allows them to compete for work with compact tractors.

Attachments for APVs can run anywhere from $500 for a plow or small trailer to $2500 for a power sweeper with a big motor. Following are some implements that are particularly useful with APVs. (See chapter 4 for a complete discussion of implements.)

**One of the most useful implements for an APV is a trailer.**

### Three-Point Adapter Trailer

A two-wheel three-point trailer allows any non-PTO powered, Category 1, three-point hitch attachment, such as a disc, harrow, or blade, to be used with a garden tractor, ATV, or UTV.

The trailer hooks to the vehicle with a pin hitch. The trailer has an electric winch that connects to the vehicle's battery via a wiring harness, and a switch in the harness raises and lowers the implement without affecting the vehicle's suspension.

In one way, this system is easier to use than the three-point hitch on a tractor because when you are finished with the implement it is a simple matter to pull the hitch pin, unplug the power, and leave the adapter trailer and implement hooked together. To use a three-point adapter, an APV should have a 400 cc or larger engine, four-wheel drive, and a minimum towing capacity of 400 pounds.

### Winch

A winch is a powered spool that typically holds at least 50 feet of cable. The winch can be bolted to a vehicle's frame or attached to a receiver hitch. Winches are available in many sizes and are rated by motor horsepower and by cable, or line, capacity. Most winches have a 12-volt motor that runs off the vehicle's battery.

A winch can be useful for tightening fence, dragging logs or rocks, or for pulling your UTV out of snow or mud. On some vehicles the winch can be used to raise and lower a plow blade. Some winches have a cabled remote control so you can position yourself out of harm's way while you operate the winch.

### Bucket

A bucket, or loader, can be attached to an APV using a universal mounting kit. It can be used to quickly move small amounts of materials like dirt, snow, gravel, and bedding and to remove manure from sheds and pens that are too small to effectively use a tractor. Once the mounting kit is installed, it can be used for several different attachments, like blades, blowers, and sweepers.

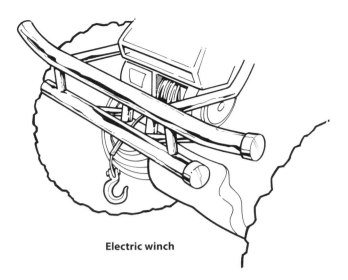

**Electric winch**

The bucket is usually raised and lowered using the vehicle's winch. The effectiveness of a bucket will be limited by the size of the vehicle using it. Trying to lift too much weight can damage the bucket and the vehicle. The most you should expect to lift with a bucket on an APV is around 300 pounds.

### Pallet Forks

Pallet forks, which can also be used with skid steers (see chapter 6), are designed to attach and detach quickly from a universal mount kit on an ATV. The forks are raised and lowered by the vehicle's winch. With some models the length of the forks and the distance between them can be adjusted.

They are handy for moving items stacked on pallets, like hay and feed, because you don't have to unstack and restack the load once you've moved it — just set the pallet where you want it and leave it with the load. Forks are also handy for moving items that don't fit easily into a loader, like long fence posts or jump rails. Leaving an empty pallet on the forks lets you use it as a flat bed for hauling just about anything you can stack on it, as long as your vehicle can safely handle the weight. Don't expect to carry more than 200 to 300 pounds on a large ATV.

### Plows and Blades

Plows and blades for APVs are available in 42-inch to 60-inch widths. They can move small amounts of material, like a one- to two-inch snowfall, quickly, but don't expect them to do the job of a heavier tractor. You'll need a kit to mount a plow on most vehicles, and the plow-mounting kit remains attached to the vehicle when the plow is removed. Blades with three-point attachments can be used with an ATV or a UTV that doesn't have a three-point hitch by utilizing a three-point hitch trailer.

## Vacuum

A lawn vacuum can be especially useful for a horse facility that wants to keep its stable yard or work areas spotless. It has its own wheeled frame or cart and is towed by a riding mower. The vacuum has a long suction hose that connects to the discharge chute of the mower's deck. A separate gasoline engine powers the vacuum to ingest just about anything the mower stirs up. The debris is collected into a bag or a bin for disposal or for use as mulch.

Most models have a handheld wand attachment on a long hose that is useful for picking up organic debris in places where a mower can't go, such as cleaning up leaves and debris from along fence lines, or vacuuming stalls and barn aisles. With some models, the handheld hose can also be used to vacuum collected debris out of the bin to deposit in places you can't reach to dump, like into a truck or over a wall. Manufacturers of lawn vacuums often sell a chipper coupled with a vacuum.

### Fire Unit

If you live in a wooded, fire-prone area, it might be good insurance to invest in a portable fire protection unit with a large plastic tank filled with a fire-fighting agent, a pump, and 50 feet of hose. Stored in such a manner that you can load it quickly into the cargo bed of your UTV or the trailer of your ATV, a fire unit might enable you to put out a small fire or at least contain it until a fire rescue unit arrives.

### Buying a Used Golf Car

One thing to be aware of when shopping for a used golf car is that some older models, mostly pre-1990, have controls that allow the vehicle to "free wheel," so if you turn off the key on a slope and get off without setting the parking brake, you could be fishing the vehicle out of the creek. Controls on most newer models prevent free wheeling.

Used golf cars are widely available and commonly range in price from $1000 to $4000, depending upon age, brand, and condition. A used golf car can provide years of use at a big savings when compared to the price of a new ATV or UTV. But will it have enough power for the chores you have in mind and will it be able to get to where you need to go?

## Buying An APV

As with purchasing a tractor, your safest bet is to buy an APV from a reputable dealer. A manufacturer's warranty (most have a standard two-year warranty) can save you money and it's good to know there's a dealer nearby who is willing to provide parts and service.

ATVs and UTVs have been on the market long enough now that there is an ample market for used vehicles. If you are buying used, follow the same guidelines outlined in chapter 3. If you are considering a used three-wheeler, be sure to check whether you can get insurance coverage for it, as many agents will not insure three-wheelers.

Buying a vehicle at auction is always risky, but if the price is right, the gamble could pay off. Do your homework and ask advice from someone knowledgeable about the type of vehicle you are considering buying. If you have Internet access, you can do a good deal of your preliminary shopping online. In a few hours you can visit the Web sites of all the major manufacturers and narrow your list of choices.

Every vehicle does something well. Find the one that best does what you want it to do. Make a list and prioritize the tasks you plan to perform with the vehicle and how many attachments you will need to accomplish those tasks. Salesmen will sometimes emphasize certain features of their products that sound impressive, but really have little bearing on your requirements. Will you ever need to go 37 mph or will 19 mph suffice?

Chances are, once you start using your new machine, you'll find more uses for it than you originally thought. But if it is too small or ill equipped for the jobs you need done most, you will end up being frustrated and disappointed and could incur costly repair bills by trying to make the vehicle do things it was not meant to do.

## TEST-DRIVE

Take at least one test-drive in each vehicle you are considering. Pay attention to how conveniently the controls are located. Does the instrument panel provide sufficient information? Is it easy to read? Is the vehicle comfortable and easy for you to handle? Do you feel safe while driving it? If at all possible, try out the vehicle on the type of terrain you will be using it on. Many dealers have varied terrain courses set up so you can get the feel driving up and down inclines and over holes and bumps. Some dealers will even haul a vehicle to your property for a test-drive.

## DECISION TIME

If you think an APV fits into your work style, use the worksheet at right to help narrow down your shopping decisions.

---

## APV Wish List Worksheet

ATV or UTV or golf car _____

Main purpose (transportation, hauling, using

implements) _____

Minimum width of openings on your property _____

_____

Required cargo weight capacity _____

Required bed or box size (UTV) _____

Required towing capacity _____

Engine size _____

Engine fuel type (gas, diesel, electric) _____

Two-stroke or four-stroke _____

2WD or 4WD or 6WD _____

Transmission preference _____

Differential lock_____

Hitch type _____

Tire type _____

ROPS (UTV)_____

Canopy (UTV, golf car) _____

Cab (UTV) _____

Number of seats (UTV, golf car) _____

Budget constraints _____

# Skid Steers and Wheel Loaders

# 6

Skid steers and wheel loaders are rugged vehicles that you have probably seen loading dirt or moving materials on construction sites. Our local farm implement dealers say they have been selling more of these versatile machines to owners of small acreages to complement or to replace a tractor. Like all vehicles, they excel at some tasks and have limitations that make them less suitable for other tasks.

# Skid Steers

A skid steer gets its name from the way it maneuvers — instead of a steering wheel that turns the wheels from side to side, a skid steer's wheels are fixed in a straight line and it uses propulsion levers to control the wheels on either side to make them go forward, stop, or go backward. To turn a skid steer, the wheels on one side lock or turn backwards while the wheels on the opposite side drive forward. This enables the vehicle to spin around within its own length, making it the handiest of all vehicles for maneuvering in tight places, like inside sheds or pens.

There are more than 10 different sizes of skid steers available, with engines ranging in horsepower from 16 to more than 90. They range in width from less than 4 feet to more than 7 feet and in length from less than 8 feet to more than 12 feet. They have a relatively low cab or rollover protective structure (most are less than 82 inches high and some are not even 70 inches), so they can get into areas such as a loafing shed where a taller tractor might not fit. An average skid steer is around the size of a utility task vehicle, but is much heavier and more powerful.

The drawback is that skid steering tears up the ground and leaves skid marks on concrete. You want to be very careful when using a skid steer on your lawn and groomed pastures and anywhere else you don't want the ground disturbed. A skid steer with normal tires can get bogged down and stuck on ground that is marshy, muddy, or covered in deep sand or snow. Tracks made of rubber, urethane, or steel, like those on military tanks, can replace or be installed right over standard skid steer tires to provide added traction and flotation. Tracks increase the effectiveness of a skid steer on soft ground. For example, they enable you to condition the footing of an arena without packing the soil like tires would.

## LOW RIDER

A skid steer's low center of gravity makes it more stable than a tractor on slopes. However, its rigid, one-piece frame, which makes it very durable, can allow one or two wheels to come up off the ground when the machine is on uneven terrain — like a block of wood with a wheel at each corner. This can result in the machine getting stuck. The rigid frame, plus the

**A skid steer is ideal for cleaning pens and loafing sheds.**

**Mounting and dismounting a skid steer is awkward and can be dangerous.**

skid steer's lack of suspension except for the air in the tires, makes this the roughest-riding and most fatiguing machine to operate for extended periods, especially over uneven terrain. Some operators have likened it to riding a bucking rocking horse.

The low ground clearance, typically between six and eight inches, predisposes a skid steer to getting caught or high-centered on protruding rocks or ridges. For jobs on rough terrain, like mowing pastures, a vehicle like a tractor that has more clearance would be a better choice. And a tractor is easier for most people to mount — to get into the cab of a skid steer, you enter from the front, which means you have to climb over the loader or whatever other attachment is hooked to the vehicle. This can be risky, especially in wet or icy weather.

## BORN TO LOAD

The skid steer's main claim to fame is its efficiency in excavating and loading bulk materials like manure and dirt. Because a skid steer is so powerful for its size, and turns so quickly, it can outperform a tractor loader hands down for these tasks. But the top ground speed of most skid steers is less than eight mph, so they are one of the slowest vehicles for

**A skid steer is very powerful and versatile for its size.**

getting from point A to point B — a good choice for tight spots, but not for long hauling.

There are many types of buckets available for skid steers, so choose the best one for the work you need to do. Some factors to consider:

◆ A utility bucket with extended side plates that hold more material in the bucket is a good choice for handling loose bedding and manure.

◆ A dirt bucket/general purpose bucket is a high-strength bucket for moving dirt and heavier materials. It has a lower capacity than a utility bucket but is stronger. Bolt-on teeth increase bucket ground penetration and reduce lip wear for digging applications.

◆ A snow bucket is the highest capacity bucket and is used for moving light materials such as snow, bedding, mulch, and fertilizers. It generally has straight sides to penetrate snow and other materials cleanly.

◆ A low profile bucket has shorter sides and back to give the best visibility of the cutting edge, which makes it convenient for grading applications, but you sacrifice bucket capacity.

◆ A grapple can be added to a bucket to hold heaped material like brush in place for carrying.

One drawback with using a machine as low and compact as a skid steer for moving dirt or manure, or even snow, is that you are very close to the dust and blowing material you are working with, whereas on a vehicle with an elevated seat, like a tractor or wheel loader, you are farther away from the material and less subject to it flying in your face. Skid steers do offer optional cab enclosures, along with heaters and air conditioners, but they can greatly limit visibility.

## IMPLEMENT OPTIONS

Skid steers are the attachment kings, with literally hundreds of different tools to choose from. They have powerful, very efficient hydraulic pumps for running hydraulic-powered implements, which most skid steer attachments are.

Quick-couple attachment systems make it easy to change from one attachment to another. A three-point adapter on the front of a skid steer enables it to use almost any three-point attachment that a tractor can. However, skid steer attachments typically cost twice as much as tractor attachments, mainly because they are all driven hydraulically, whereas many tractor attachments are run by the tractor's power take-off.

Unlike with a tractor, where you often get a sore neck from constantly twisting and turning to watch your rear attachments, skid-steer attachments connect to the front where you have a clear and easy view of where the attachment is and what it is doing.

If you have a lot of postholes to dig, a skid steer is your machine. It just can't be beat for drilling speed and efficiency. Its side-to-side, back-and-forth, up-and-down maneuverability allows you to position attachments like augers with pinpoint accuracy. And powerful downward pressure, something normal three-point tractor augers lack, enables easy digging even in extremely hard soil.

When fitted with a snowblower attachment, the skid steer's maneuverability and powerful hydraulics make it a snow-blowing monster. Most skid steer snowblowers can throw snow 45 feet. Over-the-tire tracks improve traction and efficiency in snow.

## MINI-SKID STEER

Some jobs are too big for one or two people with hand tools yet are in too small a space for a skid steer. Enter the mini-skid steer, or compact utility loader. These powerful and versatile units come in three types: walk-behind models, riding models you operate from a seat, and riding models with a rear platform for standing.

Their narrow width (many models are less than 36 inches) allow them to go through small gates and doorways. For example, a mini-skid steer on tracks would be the perfect machine for adding or removing sand in a permanent round pen where the gate is only four feet wide.

Mini-skid steers generally have an engine less than 25 hp, but have a powerful hydraulic system that

The operator's full view of skid steer attachments and the machine's strong downward force makes for fast and accurate hole drilling.

A mini-skid steer that you walk behind to operate is amazingly powerful for its size and it can get into tight places where larger machines won't fit.

A wheel loader can use many skid steer attachments and is more comfortable to operate.

can run as many attachments as a regular skid steer, including buckets, augers, pallet forks, trenchers, brooms, and stump grinders. Like standard-size skid steers, mini models can use wheels or tracks.

Controls on new models of mini-skid steers make these machines extremely easy to operate, which is one reason they are very popular with rental companies. Even if you can't justify buying a mini-skid steer, keep in mind the option of renting one for those hard jobs that your other equipment may be too large or too weak to handle.

# Compact Wheel Loader

A compact wheel loader is a small version of the giant loaders used in gravel pits to move sand, rock, and other bulk materials. Whereas large wheel loaders are basically dedicated buckets, compact wheel loaders can do many of the same jobs as a skid steer without tearing up the ground. Because they have a longer wheelbase than a skid steer, and either the rear axle oscillates or the vehicle articulates in the middle, compact wheel loaders are more comfortable to ride than a skid steer. These features also make the vehicle more stable over rough and irregular terrain. The higher ground clearance on a wheel loader makes getting stuck less likely.

Compact wheel loaders are typically larger and

more expensive than skid steers and offer fewer choices in sizes and generally less powerful hydraulic pumps. They have a higher operator station, which gives them better visibility than a skid steer and makes some of the dirtier jobs more pleasant. Using a rotary broom on a compact wheel loader, for example, is preferable, because you can see better and stay away from the dust and debris.

The elevated cab and side entrance also make it easier for the operator to get in and out, and the conventional steering wheel is more comfortable and easier for many people to operate than the levers of a skid steer.

The wheel loader's lower-horsepower engine is quieter and more fuel-efficient. Its tires incur a fraction of the wear typically experienced with skid steers. Universal quick couplers enable wheel loaders to utilize many of the same attachments that work with a skid steer.

Some newer compact wheel loaders have all-wheel-steering, where both the front and rear wheels steer. The inside tires turn at different speeds and angles than the outside tires to provide great maneuverability while minimizing ground disturbance and tire wear. For the best of both worlds, there are even machines that can lock the wheels in line and turn on a dime like a skid-steer, and then switch modes to work as an all-wheel steer machine.

# 7 Truck Features

Your rig is comprised of your towing vehicle (usually a truck) and a trailer. The selection, use, and maintenance of your truck, trailer, and hitch are of major importance. To ensure safety for you and your horse, and to protect the big investment you have in your rig, learn all you can about truck and trailer features, maintenance, and maneuvering. The first step is choosing a truck that fits the needs of your farm or ranch.

## Starting Point

There are many factors to consider when determining if a towing vehicle is suitable for pulling a horse trailer. Your choice of vehicle will depend on how much hauling you plan to do and what other uses you envision for it (commuting, ferrying the family, or taking trips). You should consider how many passengers and what type of cargo the vehicle will carry, the climate and terrain where you live, and the type of roads you typically cover (flat interstate or mountainous gravel?).

An ideal horse-hauling rig is a ¾-ton or larger truck with an aluminum gooseneck trailer. This combination will provide a safe and reliable hauling experience for you and your horse. In many cases, the engine and weight of a full-size ½-ton pickup would also be appropriate for pulling a two-horse trailer, but a ¾-ton or larger truck is capable of handling almost any farm- or ranch-hauling chore. A ¾-ton truck is needed to pull a three- or four-horse trailer. For more than four horses, for a horse trailer with living quarters, or for hauling heavy loads of hay or grain, a one-ton truck or larger would be more suitable.

Depending on the size of your operation, your needs, and your budget, make up your Dream Truck list, keeping safety in mind and taking the following characteristics into consideration.

## Towing Capacity

The towing capacity of a vehicle is very important and must be appropriate for the work. Towing capacity is determined by a number of factors including wheelbase, vehicle width, engine size, transmission, differential gear ratio, and the vehicle weight. Manufacturers determine the towing capacity for each vehicle, but these ratings are usually based on towing a static load such as a boat and trailer. Towing live weight puts greater demands on a towing vehicle than hauling fixed cargo. With horses on board, plan to tow about 25 percent less than the maximum load rating.

Overloading a tow vehicle is both illegal and unsafe. An overloaded truck requires more power to start the load moving, takes longer to come to a stop, and incurs increased wear and tear on the suspension, brakes, engine, and drive train. In general, and especially for interstate driving, the weight of a

small SUV
105"

small pickup
108"

medium SUV
112"

114"
minimum towing
vehicle wheelbase

regular pickup
116"

large SUV
119"

heavy weight SUV
131"

pickup with club cab
and 6½' bed
139"

pickup with club cab
and 8' bed
149"

pickup with crew cab
and 8' bed
165"

**The wheelbase is the distance between the centers of the front and rear axles. The longer the wheelbase, the greater the towing stability when the rig is in motion.**

loaded towing vehicle should be at least 75 percent of the weight of the loaded trailer. It's better to have more truck and not need it than to need more truck and not have it.

## WHEELBASE

The wheelbase is measured from the center of the front axle to the center of the rear axle. A longer wheelbase provides greater towing stability when the rig is in motion and decreases the likelihood of the trailer swaying.

For a short, two-horse straight-pull trailer that attaches directly to the rear of the truck, the wheelbase of the towing vehicle should be a minimum of 115 inches. Small pickups and most sport utility vehicles do not meet this criterion. If you have a two-horse trailer with a dressing room, you will need a truck with a wheelbase of at least 120 inches; for a three- or four-horse trailer, look for a vehicle with a wheelbase of 139 inches or longer. For a six- or eight-horse trailer, a towing vehicle with a wheelbase of 150 inches or longer is best.

The length of the cab and the bed primarily determine the wheelbase of your truck. From shortest to longest, cabs come in standard, Club Cab (also called extended, super, and extra cab), and Crew Cab (or four-door). The beds are available in short (6½ feet) or standard (8 feet). Flat beds, commonly seen on one-ton trucks, are 10 to 12 feet long.

# Singly vs. Dually

A truck's width, measured between the centers of the right and left tires on an axle, makes a difference in the stability of your rig on the road. The number of wheels is also a factor. Whether you select a truck with single rear wheels (SRW or singly) or dual rear wheels (DRW or dually) depends on the type of hauling and towing you plan to do, as well as personal preference. Consider handling, capacity, size, and cost.

A truck with single rear wheels can pull a two-horse trailer, but towing a larger trailer will make it more likely to sway when moving, especially with live cargo. The weight of the trailer pushes the tire sidewalls of the tow vehicle from side to side, reducing stability. A truck with a second set of heavy-duty tires on the rear axle is more stable and less susceptible to sway.

## DIFFERENCES IN HANDLING

The biggest advantage of duallies is not their towing capacity, but their stability when hauling or pulling a heavy load. Because of their wider footprint, and the added support from the sidewalls of the extra tires, they are more stable in crosswinds, when passing or being passed by 18-wheelers, and when going downhill and turning at the same time. They also sway less when going around curves and changing lanes, and they have better stopping ability. Another advantage is that in case of a rear tire blowout, you still have a remaining tire on that side to get you to a safe stop.

Having a dually, however, does not in itself guarantee safety. Safety is affected by many things like driver skill and attention, speed, road conditions, weather, and other drivers. A dually tends to have less traction, especially on slippery surfaces, because the truck's weight is spread out over a larger area, resulting in fewer pounds per square inch where the tire meets the ground. Therefore, the tires tend to "float" rather than bear down and grip. When driving through new snow or soft sand, a dually has to break six new tracks, whereas with a singly, the rear tires follow in the front tracks so that it only has to break two new tracks, which makes for easier going. Another concern is that rocks, mud, ice, and snow can wedge between the dual tires, which might damage the sidewalls of the tires, reduce traction, and damage fenders when the objects are thrown free.

## CONSIDER CAPACITY

Because of the additional set of tires and their often heavier suspension, axles, and/or larger brakes (this varies among manufacturers and across model years), duallies have a higher gross vehicle weight rating (GVWR) and gross combined weight rating (GCWR) than their singly counterparts.

Dual rear tires increase the weight-carrying capacity of the truck, allowing higher payloads and greater tongue weights. You may sacrifice a little from the GCWR and tow rating, though — a dually

**Be sure your truck has the adequate weight, length, horsepower, and towing capacity to safely pull your trailer.**

may actually be able to tow less than a similar SRW because of the increased gross vehicle weight.

Since a dually shares the load with an extra pair of wheels, it minimizes the risk of overloading the tires. Another advantage is that if you upgrade to a larger trailer that has more pin weight (tongue weight), you will have the truck to handle it.

## SIZE MATTERS

Since it is wider and sometimes higher, a dually is less maneuverable and not as easy to park as a singly. The higher stance and larger fenders can make it difficult to see small trailers and flatbeds, especially when backing. The big "hips" require you to be pretty careful in drive-through banks and restaurants. Other areas where you'll notice the increased width are going through gates, traveling across narrow bridges, negotiating narrow roads and streets, and pulling up to ATMs and gas pumps. Some duallies won't fit in certain car washes or even in your garage, if it was designed for a singly. A dually might legally be too big for your neighborhood — some covenants prohibit dually trucks, so be sure to check local rules before you go truck shopping.

**The big hips on a dually take some getting used to when entering a drive-through or when parking.**

## COST CONCERNS

A dually is more expensive both initially and down the road when you need to buy six new tires instead of four. A dually also typically gets lower fuel mileage, partly because of the lower axle ratios used for increased towing capacity and partly because of the drag of the extra wheels and the additional power it takes to drive them. In some states a dually, because of its heavier GVW, requires a license plate weight increase. Depending on what you tow, you might need to get a commercial driver's license, which means stopping at weigh stations and keeping trip logs, adding cost and time to your trip.

# Engine Issues

The towing vehicle's engine must have enough horsepower to haul the extra weight of a loaded trailer. In the mountains, figure your engine will lose two percent to four percent of its power for each 1000-foot increase in altitude. A smaller engine might save gas mileage when you are not hauling, but may result in greater repair costs due to excess strain on the engine. Assuming wheelbase and other towing requirements are met, an SUV with a 195–235-hp engine could pull a one-horse Euro-style trailer, while a full-size ½-ton truck with a 235–250-hp engine would be needed to pull a two-horse trailer. Similarly, a ¾-ton truck with a 250–300-hp engine would be suitable for a three- to four-horse trailer, and a one-ton truck with 300–345-hp diesel engine could handle a five- to six-horse trailer. (See chart on page 92.)

If you plan on owning your truck for a long time and doing a lot of long-distance hauling with it, especially in hilly or mountainous terrain, you might want to consider a diesel engine. Diesel engines can be noisy, smelly, and costly, but they are very reliable for long-distance hauling and generally last longer than gas engines. Diesel fuel can be less expensive than gasoline. Singly and dually trucks are available in either gas or diesel. (Read more about diesel engines in chapter 2.)

## Differential (Axle) Ratio

The gear ratio of the differential or "rear end" (the gear box on the rear axle) determines whether the wheels receive more of the engine's power as pulling capacity or as speed. A ratio of 4:1, for example, measures how many times the drive shaft must turn (four times) to rotate the wheels once. The higher the gear ratio, the better the torque, or pulling capacity, especially for starting, accelerating, and passing. Finding the ideal gear ratio is a compromise, however, because as the gear ratio goes up, the top end speed and the gas mileage go down. For occasional hauling, a 3.55:1 gear ratio, common in many full-size ½-ton trucks, is about optimum for a multi-purpose vehicle. If you do a lot of hauling or pull a four-horse trailer or larger, look for a rear end with a higher gear ratio. For ¾-ton trucks and heavier, a gear ratio around 4:1 is more suitable for hauling.

## Type of Transmission

An automatic transmission is easier to drive, lasts longer, generally provides a more comfortable ride for the horses, and usually has higher towing capacity than a standard transmission. A standard transmission can be tricky to operate smoothly, but usually gets somewhat better gas mileage than an automatic. However, no matter which transmission you have, when you are hauling a horse trailer, you're not really buying a truck for good gas mileage! If you choose an automatic, be sure you can lock out, or shift out of, overdrive (fourth gear). On some roads and with certain loads, the constant shifting in and out of overdrive is not only annoying, it also causes unnecessary wear on your transmission.

## Two-wheel Drive vs. Four-wheel Drive

Two-wheel drive is customarily defined as drive power delivered to the rear wheels only. Front-wheel drive is drive power delivered to the front wheels only. Four-wheel drive delivers power to all four wheels. If you live in snowy country or do the kind of off-highway driving that is often necessary on farms and ranches and at many trail ride sites, you should consider a 4WD. It adds quite a bit of extra weight to the towing vehicle's GVW, which increases the stability of the towing vehicle, but it decreases the vehicle's towing capacity and lowers gas mileage. Because a four-wheel-drive truck sits higher off the ground than a 2WD, you may need to adjust your truck's suspension or use a step-down ball mount for some trailers to ride level.

## Tire Size and Rating

Tires are critical to your truck's performance and should always be kept in good condition. Buy tires that are the correct size for your vehicle and keep them inflated with equal, optimum pressure. Tire sizes are designated by ply rating and rim size. The

**Tire size and load rating**

**Tire pressure specifications**

ply rating indicates the strength of the tire. Plies are the layers of parallel cords coated in rubber that form the body of a tire. The actual number of plies used to be marked on the sidewall of the tire, but now you are likely to find the term "load range," which rates tires from A through E, where A begins with a two-ply rating and E is the highest rating. Tires with lower ply ratings designed for passenger cars tend to flex excessively under weight and are more prone to sway.

To haul a two- or three-horse trailer, for example, a ½-ton truck should have six-ply rating, Load Range C radial tires with at least 15-inch rims. For a four-horse trailer, a truck would need eight-ply rating, Load Range D tires and 16-inch rims. For larger trucks, refer to the specifications in your owner's manual.

Tires with an open, "aggressive" tread design, such as all-terrain tires, have better traction in mud and snow, but can be noisy on the highway. This type of tread is usually not necessary with a four-wheel-drive truck. Tires with a smoother tread, such as highway or all-season tires, are quieter, run cooler, and tend to promote better gas mileage. Mud flaps will protect your trailer from gravel, mud, and dirt thrown by the rear tires of the truck.

# Putting on the Brakes

Power brakes are a must for a towing vehicle. Anti-lock brakes, which pulse rapidly to prevent locking up when used hard are ideal for a truck as they can prevent skidding. They should be on the rear wheels at least, but preferably on all four.

The brake system is designed and rated for operation of the towing vehicle itself, not the combined weight of the whole rig. Whenever you haul a trailer, you must install and use a separate brake system with a controller mounted in the truck. All states require separate trailer brakes for any trailer with a GVWR of 3000 pounds or more, which includes any horse trailer with one horse in it. These after-market brake systems are described in detail in chapter 9.

# Heavy-Duty Towing Package

Any load of more than 2000 pounds, which includes *all* horse trailers, requires the use of a heavy-duty trailer-towing package. The towing package should include a heavy-duty radiator, a heavy-duty transmission with lower gear ratio and auxiliary oil cooler, heavy-duty suspension (springs and shocks), heavy-duty battery and alternator, and a wiring harness and heavy-duty flashers. When you buy a new truck, it will only come with a hitch that has been bolted or welded to the truck's frame. You will have to select and add the ball mount and ball. Make sure the hitch has a heavy-duty receiver and is rated for the weight you plan to haul. (See chapter 8 for more information on selecting the proper hitch.)

## TRAILER WIRING HARNESS

The wiring harness typically consists of seven wires with a plug that connects to a matching socket mounted at the rear of the truck to provide electricity for the trailer's electric brakes and lights. Some plugs are flat and some are round, and the number of separate circuits, or poles, varies from three to seven. Most new trailers come with either a round six-way plug or a round seven-way plug.

There is a standard way to wire trailer plugs and tow vehicle sockets so that all trailers and vehicles using the same system are compatible. This standard utilizes color-coded wires and it works, for the most part, when hooking a new factory-wired trailer to a new factory-wired truck. You can use an adaptor to connect trailers and vehicles that have different plugs and sockets, as long as the wiring is standard on both units.

It is not uncommon, however, when hooking up used vehicles and trailers to find that the lights and brakes don't work as expected, because somewhere down the road someone rewired the plug or the socket to make it work with a different rig. And you can't just go by color when troubleshooting wiring, because color schemes vary among vehicle and trailer manufacturers. The best way to diagnose wiring problems is to use a circuit tester to find the

## Standard Wiring Schemes for Six- and Seven-wire Systems

**SIX-WIRE SYSTEM**

| | |
|---|---|
| GD (ground) | white |
| S (electric brakes) | blue |
| TM (tail, marker, license plate lights) | brown |
| LT (left turn/stop lights) | yellow |
| RT (right turn/stop lights) | green |
| A (auxiliary, extra lights, battery charger) | black |

**SEVEN-WIRE SYSTEM**

| | |
|---|---|
| 1 (ground) | white |
| 2 (electric brakes) | blue |
| 3 (tail, marker, license plate lights) | green |
| 4 (auxiliary, battery charger) | black |
| 5 (left turn/stop lights) | red |
| 6 (right turn/stop lights) | brown |
| 7 (auxiliary, backup lights) | yellow |

function of the appropriate terminals on the plug and on the socket and then rewire them to the standard.

The most common cause of trailer wiring problems is a poor ground on the tow vehicle or on the trailer. The ground wire should be connected directly to a clean portion of the frame or to a solidly attached sheet-metal part, using a screw or bolt. On most connectors, each terminal on socket and plug is labeled on the backside of the connector body with a letter or with a number.

# Calculating Weight

Your towing vehicle must be heavy enough to control the load it is pulling. To determine if it is, gather some numbers on your vehicles. See the examples on page 88 and make copies of the worksheet on page 89 for your own truck and trailer. Some numbers, such as curb weight, maximum tow weight rating, and payload, are available from the manufacturer or dealer. You can get others, such as gross vehicle weight and gross cargo weight by using a scale and calculations. Commercial weigh scales for weighing semi-tractor/trailers can be found at your feed mill, gravel pit, or at highway truck checks. Plan to pay a nominal fee for the weight ticket.

According to U.S. towing guidelines, tongue weight, the amount of downward weight the tongue puts on the hitch, is 10 to 15 percent of the total trailer weight. Tongue weight is especially important to consider when towing with a car or truck at speeds faster than 20 mph, because too little tongue weight (if it is caused by having the bulk of the load too far back on the trailer) can cause the trailer to fishtail or sway uncontrollably.

## CURB WEIGHT

Curb weight is the weight of the empty vehicle, and the best way to determine it is to weigh your empty truck on a commercial scale. Actual curb weight usually includes a driver (but no passengers), a full tank of gas and other fluids, and standard equipment such as spare tire, jack, and other items.

To estimate your truck's curb weight without weighing it, first find the manufacturer's curb weight for your truck. Most truck dealers can provide this information (with plus or minus 3 percent accuracy). In order to give you the most accurate figure, the dealer will need the year, model number, engine, transmission, cab style, length of bed, and whether the truck has two-wheel drive or four-wheel drive. The manufacturer's curb weight figure represents a vehicle with standard equipment. For the ½-ton truck in the example, the dealer provided the figure of 5000 pounds.

To calculate actual curb weight from the truck manufacturer's base figure, add the weight of any nonstandard equipment you have installed, such as a camper shell, running boards, large mirrors, heavier bumper, or pickup bed liner or tonneau cover. This total will be an estimate of your truck's actual curb weight. The actual curb weight of the ½-ton truck in the example, as determined by a scale, is 5300 pounds, which reflects 300 pounds of add-ons.

## GROSS VEHICLE WEIGHT RATING

Gross vehicle weight rating is the maximum allowable weight of the truck and its payload as established by the manufacturer. This includes trailer tongue weight. For safety, the towing vehicle with cargo should never weigh more than the GVWR. In addition to the GVWR, some manufacturers will provide a maximum tongue weight rating (MTWR).

The GVWR is listed on the vehicle identification plate, often located on the driver's door latch pillar. The GVWR also appears on the window sticker of a new vehicle and in other paperwork and brochures. The GVWR of the example ½-ton truck in the weight chart on page 88 is 6400 pounds.

Payload is the maximum allowable weight of cargo, passengers, and tongue weight that the truck is designed to carry. This figure is available from the vehicle manufacturer or dealer.

*GVWR – curb weight = payload*

6400 – 5000 = 1400 for the example ½-ton truck

If you can't find the standard curb weight of your vehicle (and don't have access to a scale), you can calculate the standard curb weight if you know the GVWR and payload.

*GVWR – payload = standard curb weight*

6400 – 1400 = 5000 for the example ½-ton truck

If you are trying to estimate the *actual* curb weight using this method, don't forget to add the weight of nonstandard equipment that you have added to the truck. For the example ½-ton truck, since the actual curb weight is 5300 pounds (as determined by scale), and the GVWR is fixed at 6400 pounds, the actual payload (additional weight it can carry) is reduced to 1100 pounds.

## GROSS VEHICLE WEIGHT

Gross vehicle weight is the actual weight of your loaded truck, so it changes according to what is in the truck. It is the actual curb weight plus the weight of any passengers (other than the driver who is already figured in the curb weight) and the weight of any cargo in the truck. The GVW should never exceed the GVWR. If you are pulling a trailer, the GVW also includes the tongue weight of the trailer (See page 88 as well as chapters 8 and 9). The GVW is best determined by weighing, but it can also be estimated by calculation. The result should be compared to the GVWR.

## Vehicle Identification Plate (VIP)

The VIP, located inside the driver's door of a truck and on the side or tongue of a trailer, should list:

❏ The manufacturer's name, location, and date of manufacture

❏ GVWR (gross vehicle weight rating)

❏ Tongue weight (not all manufacturers)

❏ GAWR (gross axle weight rating) for both the front and rear axles

❏ Tire size

❏ Rim size

❏ Recommended tire pressure (per square inch) cold

❏ VIN (vehicle identification number)

❏ Type of vehicle (such as truck) and whether single or dual rear wheels

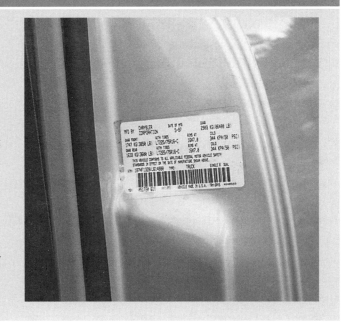

For the example truck, the GVW is 5900 pounds with an empty steel trailer attached. The GVWR for this vehicle is 6400 pounds, so when two horses are put in the trailer, the GVW increases to 6260 pounds due to the increase in tongue weight. This is quite close to the 6400 pounds GVWR of the truck, and leaves only 140 pounds for passengers and cargo. (See other truck-trailer combinations on page 88.)

The most accurate determination of tow vehicle's GVW is obtained by weighing the loaded truck with loaded trailer attached on a commercial scale. Only the truck's wheels should be on the scale for

**Your truck's actual curb weight will be higher than the manufacturer's if you have add-ons like this tonneau cover.**

## Weight Definitions for Trucks and Trailers

To discuss truck weights without talking about trailer weights would be incomplete, so they are both covered here.

### CURB WEIGHT (CW)

*Truck:* the weight of an empty vehicle plus gas and driver

*Trailer:* the weight of an empty trailer not hitched to a vehicle

### PAYLOAD

*Truck:* maximum allowable weight that the truck is designed to carry, per manufacturer

### GROSS VEHICLE WEIGHT RATING (GVWR)

*Truck:* maximum allowable weight of the truck plus its cargo, per manufacturer

*Trailer:* maximum allowable weight of the trailer plus its cargo, per manufacturer (sometimes called the loaded trailer weight rating or LTWR)

### GROSS VEHICLE WEIGHT (GVW)

*Truck:* the actual weight of the truck, which includes all cargo, passengers, and the trailer tongue weight; varies with each load

*Trailer:* the actual weight of the trailer, which includes weight of the horses, tack, feed, water, and other cargo minus the tongue weight (sometimes called the loaded trailer weight or LTW); varies with each load

### GROSS COMBINATION WEIGHT RATING (GCWR)

*Rig:* the maximum allowable total weight (per truck manufacturer) of the loaded truck (GVWR) plus the loaded trailer (GVWR)

### GROSS COMBINATION WEIGHT (GCW)

*Rig:* the actual weight (GVW) of the loaded truck plus the actual weight (GVW) of the loaded trailer; varies with each load

### TONGUE WEIGHT

The amount of the trailer's weight that is transferred to the truck via the tongue or gooseneck; usually provided by the trailer manufacturer as a percentage of the GVW of the trailer

### MAXIMUM TOW WEIGHT RATING (MTWR)

*Truck:* the maximum allowable towing capacity of a vehicle. Some manufacturers will also provide a maximum tongue weight rating which is usually approximately 10 percent of the MTWR.

this measurement. The trailer tongue weight will be a part of the GVW of the truck. The GVW of the example ½-ton truck with Trailer A is 6260 pounds; with Trailer B, the GVW is 5900 pounds.

## TRAILER TONGUE WEIGHT

Tongue weight refers to the amount of the trailer's weight that is being transferred from the trailer to the truck. While a straight-pull rig transfers tongue weight to only the rear axle of the towing vehicle, a properly mounted gooseneck or fifth-wheel hitch results in a portion of the tongue weight of the trailer being distributed to the front wheels of the towing vehicle as well.

Straight-pull trailers usually transfer from 10 to 15 percent of their weight as tongue weight to the rear axle of the towing vehicle, while gooseneck and fifth-wheel trailers typically transfer about 25 percent of their weight to the towing vehicle. Euro-style trailers are said to transfer only 3.75 percent of their weight to the rear axle of the tow vehicle.

Too much tongue weight can damage the suspension and/or drive train. With a straight-pull, too much tongue weight on the rear end of the towing vehicle can lift the front wheels and seriously effect steering and braking. A trailer that is heavily loaded in the front or is unbalanced could deliver too much tongue weight to the truck. However, when a trailer is loaded heavily at the rear, the tongue of the trailer may elevate and lift the truck up at the ball, reducing

## Weight Chart Example

**EXAMPLE TRUCK:** ½-TON PICKUP

| | |
|---|---|
| Curb weight, per manufacturer | 5000 |
| Payload per manufacturer | 1400 |
| GVWR per manufacturer | 6400 |
| GVW with add-ons, on scale | 5300 |
| GVW with empty steel trailer attached, on scale | 5900 |
| GVW with empty aluminum trailer attached, on scale | 5630 |
| GVW with loaded steel trailer attached | 6260 |
| GVW with loaded aluminum trailer attached, on scale | 5900 |
| MTWR (per manufacturer) | 7300 |
| GCWR (per manufacturer) | 12,500 |
| GCW with loaded steel trailer | 11,300 |
| GCW with loaded aluminum trailer | 9940 |

**EXAMPLE TRAILER A:** TWO-HORSE STEEL SLANT-LOAD WITH DRESSING ROOM

| | |
|---|---|
| Curb weight per manufacturer, standard features | 3600 |
| Curb weight with options, on scale | 3860 |
| Tongue weight, per trailer manufacturer | 540 |

| | |
|---|---|
| Tongue weight percentage, per trailer manufacturer | 15% |
| Tongue weight, empty, on scale | 600 |
| Tongue weight percentage, actual | 16% |
| Tongue weight, loaded, by calculation | 960 |
| GVW (3860-pound trailer, 2000 pounds horses, 140 pounds tack, feed, water) | 6000 |
| GVWR | 7000 |

**EXAMPLE TRAILER B:** TWO-HORSE ALUMINUM STRAIGHT-LOAD WITH DRESSING ROOM

| | |
|---|---|
| Curb weight, per manufacturer | 2500 |
| Tongue weight, empty, per trailer manufacturer | 300 |
| Tongue weight percentage, per trailer manufacturer | 12% |
| Tongue weight, empty, on scale | 330 |
| Tongue weight percentage, actual | 13% |
| Tongue weight, loaded, by calculation | 600 |
| GVW (2500-pound trailer, 2000 pounds horses, 140 pounds tack, feed, water) | 4640 |
| GVWR | 7000 |

rear wheel traction. This type of imbalance is most likely to cause dangerous sway and fishtailing. The truck is no longer in control because the heavy trailer is like "the tail wagging the dog."

## DETERMINING TONGUE WEIGHT

There are three ways to determine the tongue weight of your trailer.

**Ask.** Your dealer will be able to give you an estimate for a standard trailer. The dealer figures supplied for both of the sample trailers were somewhat lower than the weight obtained from commercial scales. This discrepancy could be due to options added by the manufacturer (side wall mats, spare tire, saddle racks, for example), which will affect overall weight and tongue weight, especially if the add-ons are in front of the trailer's axles. The difference between manufacturer figures and actual scale weight could also be a scale-calibration issue, making a difference up to 100 pounds measured on scales designed for semi trucks that weigh many tons.

**Weigh.** Using a commercial scale, weigh your truck; then weigh it with the trailer attached. First drive your truck alone onto a scale. The example ½-ton truck weighs 5300 pounds. Then attach your trailer, checking that the tongue, coupler, and receiver are horizontal and level when the trailer is fully hitched. Drive onto a scale, but make sure that only the truck's tires are on the scale. The example ½-ton truck with Trailer A weighs 5900 pounds. To determine the tongue weight, subtract the truck's known GVW from the GVW with trailer attached. In the example, 5900–5300 = 600 pounds for the tongue weight of the empty steel trailer. For Trailer B, 5630–5300 = 330 pounds. You can use this method with an empty or loaded trailer. But if you do this for a loaded horse trailer, weigh your truck separately (without trailer attached) to find its actual GVW and then come back with the loaded trailer to find the "tongue" weight. *Note:* You should never unhitch a loaded horse trailer.

**Calculate.** You can figure the tongue weight for an empty trailer by first taking the GVW of the

## Truck and Trailer Worksheet

Use this worksheet to record the weight of your current truck and trailer and for comparison, any new trailer you are considering purchasing.

**Your truck** _____

Curb weight per mfr. _____

Payload per mfr. _____

GVWR per mfr. _____

GVW loaded on scale _____

GVWR – GVW (balance available for trailer tongue weight and more passengers or cargo) _____

GCWR per mfr. _____

**Your current trailer** _____

Curb weight per mfr. _____ on scale _____

Tongue weight per mfr. _____ on scale _____

Tongue weight % per mfr. _____ on scale _____

Tongue weight, loaded, by calculation _____

  OK with truck? _____

GVWR per mfr. _____

GVW on scale _____

  OK with GVWR above? _____

GCW _____

  OK with truck GCWR? _____

**Another Trailer** _____

Curb weight per mfr. _____ on scale _____

Tongue weight per mfr. _____ on scale _____

Tongue weight % per mfr. _____ on scale _____

Tongue weight, loaded, by calculation _____

  OK with truck? _____

GVWR per mfr. _____

GVW on scale _____

  OK with GVWR above? _____

GCW _____

  OK with truck GCWR? _____

truck with trailer attached. With the truck still on the scale, block the empty trailer securely, unhitch it, and make sure the coupler is not touching the ball of the receiver and the trailer jack is not on the scale. Note the weight of your towing vehicle now, making sure all passengers and cargo have remained in the vehicle for both measurements (5300 pounds for example vehicle). Subtract the second figure from the first figure. For Trailer A, 5900 − 5300 = 600 pounds of tongue weight. For Trailer B, 5630 − 5300 = 330 pounds. This is the tongue weight of the empty trailer.

## TONGUE WEIGHT PERCENTAGE

When your trailer is hitched to your truck, you add the tongue weight to the truck's GVW and subtract it from the trailer's GVW. In essence, the trailer is giving that weight to the truck to carry. So when Trailer A is attached, the truck's GVW is 6260 pounds and the truck is towing (and the trailer is carrying) 5040 pounds (6000 − 960 = 5040 pounds). When Trailer B is attached, the truck's GVW is 5900 pounds and the truck is towing (and the trailer is carrying) 4040 pounds (4640 − 600 = 4040 pounds).

To determine the percentage of the trailer's weight that is being transferred to the truck via the tongue, divide the tongue weight by the trailer weight. For Trailer A, the tongue weight (600 pounds) divided by the trailer weight (3860 pounds) results in a tongue weight of 16 percent. For Trailer B, the tongue weight (330 pounds) divided by the trailer weight (4640 pounds) is 13 percent.

### Commercial Driver's License

Since 1992, federal law has required individual states to issue a commercial driver's license (CDL) to commercial drivers who operate a CDL-class vehicle, including a truck with a GVWR of 26,001 pounds or more, or a trailer with a GVWR of 10,001 pounds or more, if the GCWR of the trailer and the truck equals 26,001 pounds or more. Recreational truck/trailer units are generally exempt if they are not being used for commercial purposes.

Theoretically, as long as the load is fairly evenly balanced, the tongue weight percentage should be approximately the same whether the trailer is loaded or not, though the actual transfer of weight will be greater from a loaded trailer than from an empty one. Once you have calculated the empty trailer tongue weight and percentage, you can use that percentage to calculate an estimate of the loaded trailer tongue weight.

For Trailer A, if the loaded trailer weight is 6000 (3860 pounds trailer, 2000 pounds horses, 140 pounds tack, water, and feed) then the weight transferred to the truck is 960 pounds, or 16 percent. When added to the example truck's GVW of 5300 pounds, the resulting 6260 pounds is close to the vehicle's GVWR of 6400 pounds. Trailer B's loaded trailer tongue weight is 13 percent of 4640, or 600 pounds. This would make the GVW of the truck 5300 + 600 = 5900 pounds, which is 500 pounds under the GVWR, a safer margin.

## GROSS COMBINATION WEIGHT RATING

Gross combination weight rating (GCWR) is the maximum allowable weight of the entire rig: the loaded vehicle plus the loaded trailer. You can get this figure from the truck's manufacturer. The GCWR helps you calculate how much weight your truck can safely and legally carry and tow. The GCWR of the example ½-ton truck is 12,500 pounds.

## GROSS COMBINATION WEIGHT

Gross combination weight (GCW) is the GVW (actual weight) of the fully loaded vehicle plus the GVW (actual weight) of the fully loaded trailer. You can calculate an estimate of the GCW, but again, the most accurate determination is to weigh your entire loaded truck and trailer on a commercial scale. To obtain the GCW of your entire loaded rig, make sure all tires of both the truck and trailer are completely on the scale. The GCW of the example truck with loaded Trailer A is 11,300 pounds and with loaded Trailer B, it is 9940 pounds. The GCW should never exceed the GCWR.

# General Guidelines for Safe Vehicle/Trailer Combinations

The designations ½-ton, ¾-ton, and one-ton indicate a vehicle's approximate load carrying capabilities. As you go up in numbers, the vehicle becomes capable of carrying more weight, rides higher off the ground, and is more heavy duty, but provides a stiffer ride and gets lower mileage. You can use the following rules of thumb to estimate safe towing capacity and to get an idea of the relationship between the towing vehicle and the loaded trailer. However, be sure to actually weigh and calculate your rig's components to be certain and safe.

Usually a towing vehicle can safely pull approximately 1⅓ times its curb weight. For the example truck with a manufacturer's curb weight of 5000 pounds, 1⅓ times 5000 is 6650 pounds. The maximum trailer weight rating (MTWR) for this vehicle is 7300 pounds. It is sometimes difficult to obtain a vehicle's MTWR from the manufacturer, so this rule gives you a safe estimate.

For interstate driving, the GVW of the truck should be at least 75 percent the GVW of the loaded trailer. For example, the GVW of Trailer A is 6000 pounds; Trailer B's is 4640 pounds. The towing vehicle should weigh at least 4500 pounds to haul Trailer A and 3480 pounds to haul Trailer B. With a GVW of 5300 pounds, the example ½-ton truck has adequate weight. Here are a couple of safe combinations of truck and trailer to consider.

**Combination 1.** Use a ¾-ton truck to pull a steel trailer with a dressing room. This would be the best option because of the GVW and tongue weight of this trailer. The GVWR of a ¾-ton truck is higher than that of a ½-ton truck, so the tongue weight should pose no problem. As the GCWR of the ¾-ton truck is also higher, you will have a greater margin of safety and more cargo capacity.

**Combination 2.** Use a ½-ton truck to pull a lighter trailer, such as a one- or two-horse Euro-style trailer or a two-horse steel trailer without a dressing room (2500- to 3000-pound curb weight) or a two-horse aluminum trailer with a dressing room (2500-pound CW). The lower trailer GVW and lower tongue weight of either of these choices will keep the GVW of the truck and the GCW safely below the ratings.

## HAULING WITH AN SUV

Using a small to medium sport utility vehicle or small pickup for towing a conventional horse trailer is not a safe option for several reasons. Most small SUVs and small pickups have tow ratings of less than 5000 pounds. And for some of those vehicles (with good

**The actual GVW of the truck includes all cargo, passengers, and trailer tongue weight. This can be estimated by calculation or measured with all four tires of the truck on a scale and with the loaded trailer attached.**

reason) you can't buy a hitch rated high enough to pull 5000 pounds.

Even a medium SUV isn't a good choice for hauling a horse trailer. First of all, the wheelbase is less than 112 inches, which is below the safe minimum of 115 inches. With an automatic transmission, a heavy-duty towing package, and a high-performance axle, the towing capacity is 5700 pounds, which technically means it could pull a light two-horse trailer (no dressing room) with two average-sized horses. However, the curb weight of the vehicle is 4500 pounds and the GVWR is 5000 pounds, leaving only 500 pounds for tongue weight, passengers, and cargo. Even though the GCWR is nearly 10,000 pounds, just about any loaded two-horse trailer would transfer too much tongue weight for the vehicle's GVWR and the combined weights would be higher than the GCWR. A better match would be a one- or two-horse Euro-style trailer, which delivers only 3.75 percent tongue weight to the towing vehicle.

| TOW VEHICLE | APPROPRIATE SIZE TRAILER |
|---|---|
| Regular size SUV | One-horse Euro-style trailer (2600 pounds loaded) or two-horse Euro-style trailer (4400 pounds fully loaded) |
| Full-size ½-ton truck or large SUV | Two-horse straight-pull trailer with no dressing room (up to 6000 pounds fully loaded) |
| ¾-ton truck | Two-horse straight-pull trailer with dressing room or three-to four-horse gooseneck trailer (up to 10,000 pounds fully loaded) |
| One-ton dually | Four- to six-horse gooseneck trailer with living quarters (up to 20,000 pounds fully loaded) |

If you must use an SUV, purchase a large one with a 119-inch wheelbase. If outfitted with automatic transmission, a V-8 engine, and proper towing package, it would have a 12,500-pound GCWR and a trailer towing capacity of 7000 pounds. It could safely tow a two-horse trailer, especially if you use a weight-distribution hitch.

# The Hitch

**Y**our trailer hitch is a critical component of your rig, and you rely on it for the safety of you and your horses. There are three types of hitches used for connecting trailers to towing vehicles: the straight-pull, the gooseneck, and the fifth-wheel. There are advantages and disadvantages to each type, and which you choose will depend on the size and type of trailer you want to pull, the type of towing vehicle you have, and your personal preferences.

Every hitch is a two-part unit, consisting of a coupler attached to the tongue or bunk of the trailer and a matching receiver bolted or welded to the frame of the towing vehicle. A straight-pull hitch is accessible at the back of the vehicle and a gooseneck or fifth-wheel hitch is located near the center of the bed.

A new trailer comes fitted with a coupler, but a new truck does not automatically come with a receiver installed — you will have to specify one when ordering the truck or add one after purchase. You will also need to buy a ball and possibly a ball mount that is rated to use with the hitch and that will fit the coupler.

# Straight-pull Hitch

A straight-pull hitch, also called a frame hitch, is installed on the frame of the tow vehicle, generally below the rear bumper. The hitch has a ball mount and ball fastened to it, and the trailer has a coupler that sets over the ball and locks onto it. There are two types of frame hitches: the fixed hitch and the receiver hitch. Most frame hitches are rear-mounted but they can also be installed on the front of a vehicle to make parking trailers easier.

## FIXED HITCH

A fixed hitch, also called a fixed tongue, is permanently attached to the vehicle's frame, so it remains in place when the trailer is unhitched. The big disadvantage to this is that the hitch is a "shin buster" when the trailer is disconnected.

A fixed hitch can be used only as a weight-carrying hitch, not as a weight-distributing hitch, which limits your vehicle's towing capacity. This type of hitch has been replaced for the most part by the receiver hitch, which is more convenient and more versatile.

## RECEIVER HITCH

A receiver hitch has two parts. The receiver part attaches to the vehicle's frame like a fixed hitch, but the ball mount and ball are removable. This type of hitch has a 1¼-inch or 2-inch square steel tube into which a matching steel shaft that holds the ball is inserted and locked in place with a removable pin and clip. The shaft is also called the ball mount or drawbar, and a ball of any size can be bolted to it. A receiver hitch can be used as a weight-carrying hitch and, if sufficiently rated, a weight-distributing hitch. The ball mount can be removed when the hitch is not needed. The end of the receiver hitch is then about flush with the end of the vehicle where you are unlikely to bump into it when you walk behind your truck.

The ball mount must be set at the proper height to allow the trailer floor and the tongue to be level when a loaded trailer is attached to a truck. Ball mounts of various configurations — straight, stepped upward or downward, or adjustable — can be matched with different trailers to ensure that the trailer rides level. A straight-pull trailer should carry 85 to 90 percent

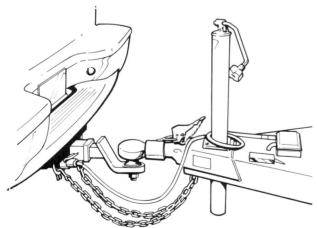

**Straight-pull hitch**

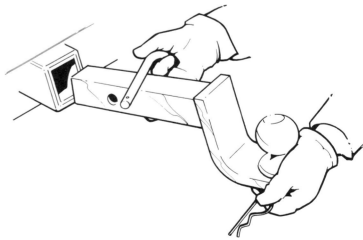

**The ball mount can be removed from a receiver hitch when not in use.**

### Ball Mounts

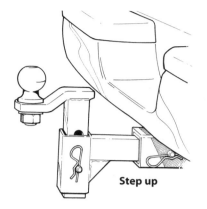

**Step up**

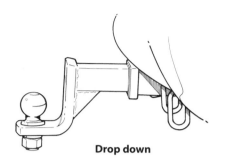

**Drop down**

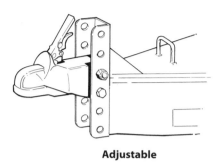

**Adjustable**

of its weight on its wheels and transfer only 10 to 15 percent to the truck via the hitch.

If the coupling is too high, the floor of the trailer will slant downhill toward the rear of the trailer. The excess weight behind the rear axle of the trailer can actually lift up on the rear of the towing vehicle and cause vehicle and trailer sway, as well as making the

ride uncomfortable for the horses. If the coupling is too low, the floor of the trailer will slant downhill toward the front of the trailer and there will be excess wear on the ball and excess tongue weight on your towing vehicle.

The drop height of ball mounts (the amount the ball is lower than the receiver) varies from 1½ inches

## The So-called Bumper Pull

A straight-pull trailer is often incorrectly called a bumper pull trailer, even by manufacturers. Years ago, when vehicles and their bumpers were made of very heavy materials, some full-size autos that had a long wheelbase, low center of gravity, heavy curb weight, and V-8 engine could tow substantial trailers attached to their bumpers. Today, even truck bumpers are typically lightweight and could easily be damaged or come loose if a horse trailer were attached to them. Most bumper hitches on light trucks and sport utility vehicles are rated for very light duty, less than 3500 pounds, and are not suitable for pulling a horse trailer.

So even if your trailer is described as a "bumper pull," *do not* attach it to your vehicle's bumper. Even if the bumper is rated to pull the GVW of your loaded trailer and you make sure the trailer is level, using the bumper is not safe. For one thing, even if you have a heavy-duty truck with a heavy-duty bumper, the trailer will follow so closely to the truck that it is

likely to hit it when turning. A frame hitch with ball mount locates the point of attachment of the trailer low enough to keep the trailer level and about eight inches behind the bumper, resulting in better clearance on tight corners.

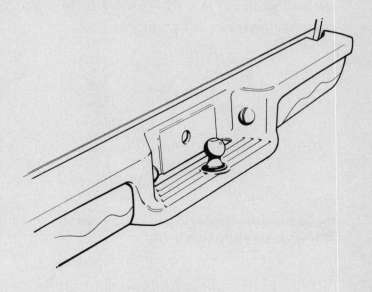

A level truck and trailer

This trailer's weight is imbalanced to the rear resulting in an uphill ride for the horse, more weight borne by the trailer's rear axle, and the trailer lifting up on the rear of the truck which can cause fishtailing and loss of traction.

to 8 inches to accommodate the various heights of truck and trailer. Ball mounts can also be turned over and the ball bolted to the other side if the trailer's coupler is higher than the hitch. Adjustable ball mounts, rated up to 10,000 pounds and with 11½ inches of vertical adjustment, are available if you tow trailers of different heights. Ball mounts should be removed when not in use as they are easily stolen and a common cause of leg injuries.

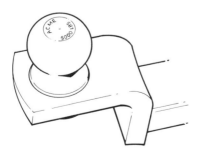

The weight rating is stamped on the ball and should be sufficient for the weight being towed.

Each component of the hitch must be rated high enough for the gross vehicle weight of the trailer and the tongue weight of the trailer. There should be a stamp or decal on each item listing its weight rating. Straight-pull and gooseneck balls should be rated sufficiently for the hitch and for the trailer's GVW. If the stamped size is illegible, you can measure the diameter of the ball using a pair of vernier calipers. The shaft, or shank, of the ball should be the same size as the hole in the ball mount.

## Weight-distribution Hitch

A weight-distribution hitch is a substitute for, or an add-on to, a receiver hitch that can increase your vehicle's towing capacity up to 14,000 pounds and its tongue weight capacity up to 1400 pounds. A weight-distribution hitch transfers tongue weight to the front axle of the tow rig and to the trailer axles. This redistribution of weight increases the stability of the whole rig.

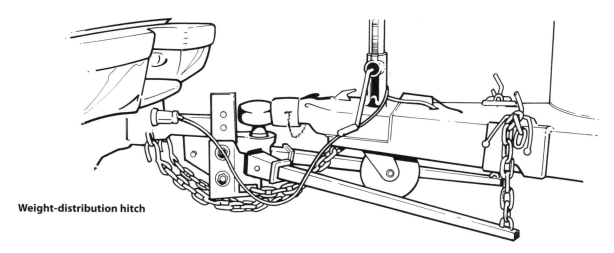

**Weight-distribution hitch**

Although tongue weight is necessary for the truck to control the trailer, when the tongue weight is too heavy, weight is transferred from the truck's front axle to its rear axle, increasing the rear loading. For tongue weights greater than 700 pounds, the truck can become unbalanced enough to affect driving control. If a truck is light in the front end, the steering, turning, and braking will be negatively affected and the truck will be less able to control sway. You can't visually assess if tongue weight is appropriate, because often the springs are strong enough so the truck looks fine (not raised in front or down in back) even though the rig is not balanced.

A weight-distribution hitch has two steel bars, called spring arms, which are part of the ball mount and run parallel underneath the A-frame tongue of the trailer. Two cuffs, or brackets, are attached to the trailer frame, one on each side, in line with the ends of the arms. When you hook up the trailer, chains at the ends of the spring arms connect to the cuffs. On each cuff, a type of rocker lever tightens the chains and basically lifts the truck and trailer at the hitch point, transferring weight to the front axle of truck and rear axle of the trailer. The trailer is now being supported or towed from three points (the ball and two cuffs), as opposed to just one point (the ball) in a normal hitch assembly.

## HITCH CLASSIFICATION

Frame hitches are rated according to the weight they can tow. Generally, the tongue weight rating of a hitch is 10 percent of the hitch rating, so a hitch rated to carry 10,000 pounds is rated to carry 1000 pounds of tongue weight. Since there is no standardization among manufacturers as to what constitutes a Class III or Class IV hitch, only use class designation as an approximation. Find out the actual weight rating of a particular hitch and know what it means before you install the hitch.

There are two weights, for example 5000/10,000, listed on the frame portion of the hitch. The lower number refers to the hitch's maximum capacity in a

## Common Ball Sizes

♦ 1⅞-inch ball with ¾-inch or 1-inch shank; 2000-pound capacity

♦ 2-inch ball with ¾-inch shank; 3500- or 5000-pound capacity

♦ 2-inch ball with 1-inch shank; 5000- or 6000-pound capacity

♦ 2-inch ball with 1¼-inch shank; 6000-, 7500-, or 8000-pound capacity

♦ 2⁵⁄₁₆-inch ball with 1-inch shank; 6000-pound capacity

♦ 2⁵⁄₁₆-inch ball with 1¼-inch shank; 10,000-pound capacity

♦ 2⁵⁄₁₆-inch ball with 1¼-inch shank (heat treated); 14,000-, 24,000-, or 30,000-pound capacity

♦ 3-inch ball with 2-inch shank (gooseneck only); 30,000-pound capacity weight-carrying situation where a weight distribution hitch is not being used. A 5000-pound maximum trailer weight rating (and corresponding 500-pound maximum tongue weight capacity) would be inadequate for many two-horse trailers with dressing rooms. The higher figure indicates the capacity of the hitch with a weight distribution system added to the truck and trailer. If you use a weight-distribution hitch, you'd be able to tow up to 10,000 pounds and carry up to a 1000-pound tongue weight.

For almost any straight-pull trailer, it is best to use a 7500/12,500 or a 10,000/14,000 hitch if your

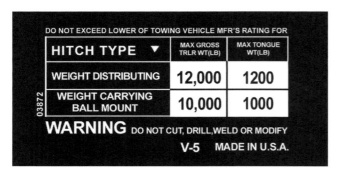

**Decal on a frame hitch**

## Trailer Hitch Estimated Capacity

**Class I:** Up to 2000 pounds gross trailer weight and 200 pounds tongue weight; 1¼-inch receiver.

**Class II:** Up to 3500 pounds gross weight and 350 pounds tongue weight; 1¼-inch receiver.

**Class III:** Up to 5000 pounds gross weight and 500 pounds tongue weight. (Sometimes refers to a hitch with a 2-inch receiver, regardless of rating.)

**Class IV:** Up to 10,000 pounds gross trailer weight and 1000–1400 pounds tongue weight; 2-inch receiver. (Frequently, any hitch with a capacity greater than 5000 pounds gross weight is referred to as a Class IV.)

**Class V:** Capacity greater than 10,000 pounds gross trailer weight and 1000–1200 pounds tongue weight; usually a 2½-inch receiver or, in some special cases, a 2-inch receiver.

tow vehicle can handle it. Be sure the ball mount and ball you use are rated for those weights as well.

**A word of warning:** Don't assume that the heavy-duty towing package that comes on your new truck will be adequate for the truck's maximum trailer weight rating. There's a good chance the manufacturer or dealer will install a 5000/10,000 hitch and tell you it's good for 10,000 pounds, letting you think you can hook up your loaded two-horse trailer (7000 pounds) and take off. Not true! In order to safely tow more than 5000 pounds with that hitch, you'd need to add a weight distribution hitch.

Pay attention to the kind of hitch that comes with your truck, because some manufacturers don't even offer a hitch that is adequate for their ½-ton trucks. You might need to buy an after-market heavy-duty hitch if you want to safely pull a loaded two-horse trailer without using a weight distribution hitch. The lowest number of the hitch's maximum trailer weight rating must be at least as high as the truck's MTWR and higher than the loaded trailer weight you plan to pull, unless you plan to use a weight-distribution hitch every time you tow.

Generally, Class I and II receiver hitches can only be used as weight-carrying hitches. Class III and IV hitches can be used as weight-carrying hitches and also as weight-distributing hitches when a weight-distribution system is added.

## SWAY CONTROLS

If your rig sways when you are driving it, you may need to consider installing sway controls. These can be a separate add-on or a part of a weight-distributing hitch, but before installing them, be sure you actually need them. Sometimes factors that can be corrected are causing the sway.

First check that your tongue weight is in a normal range for your rig. If it is high, check to see if the receiver on your truck is adequately rated. You could also consider using a weight-distributing hitch. If the tongue weight is too low, the trailer will sway the truck, so you need to increase tongue weight by moving your trailer load forward. Then check your tires. Low air pressure in either the truck or trailer tires can result in sway. Tires that are rated too low for your load could also cause a problem. Make sure that the trailer is riding level, with the coupling neither too high nor too low.

Once you have evaluated the various causes of trailer sway, you might consider auxiliary anti-sway devices. Don't confuse "sway controls" with "sway bars" that attach to the tow vehicle's suspension and have no effect on trailer sway. There are two basic types of sway controls: a friction sway control and a cam sway control. A friction sway-control device stiffens the coupling between the tow vehicle and trailer by means of a two-piece rod, somewhat like a shock absorber, mounted between the ball mount and the trailer's A-frame. Friction sway controls don't prevent sway; they simply resist sway forces once they have started. The degree of stiffening or friction can often be adjusted to suit various trailer weights and towing conditions.

Unlike a friction control, a cam sway control prevents sway in the first place. When towing in a straight line, cams on either side of the trailer lock into position to create a rigid connection between the tow vehicle and the trailer that minimizes sway caused by high crosswinds or passing vehicles. During normal cornering, the cams automatically slide out of their locked position to permit full-radius turns. If the maneuver is short and abrupt, like a sudden swerve or a wheel dropping off the road, the cams lock to help the tow vehicle retain control.

## Goosenecks and Fifth-wheels

A gooseneck trailer is immediately recognizable because it extends over the bed of the truck. It attaches to the frame of the truck slightly ahead of the rear axle with either a gooseneck hitch or a fifth-wheel hitch. Both the gooseneck hitch and the fifth-wheel hitch have a hitch component on the truck and a coupler component that is part of the trailer. The truck portion of both types is secured to the tow vehicle by rails bolted between the vehicle's right and left frame members underneath the bed. The hitch is then fastened to the rails through the bed. Because the hitch is mounted in the truck bed two to four inches ahead of the rear axle, approximately 95–98 percent of the tongue weight rests on the rear axle, with the remainder of the tongue weight carried by the front axle.

**Gooseneck trailer**

Both gooseneck and fifth-wheel hitches have a vertical "tongue" portion on the trailer that extends down from the bunk and holds the coupler. With some newer trucks and trailers, a gooseneck coupler can be adapted for use with a fifth-wheel hitch, and a fifth-wheel coupler can be adapted to use with a gooseneck hitch.

## GOOSENECK HITCH

With a gooseneck trailer and hitch, you can tow far more weight than with a similar straight-pull trailer. The coupler and ball of a gooseneck hitch are typically rated at 25,000 to 30,000 pounds tow capacity. On the truck, sometimes all you can see of a gooseneck hitch is a 2⁵/₁₆ ball (similar to that used with a straight-pull trailer) mounted in the center of the truck bed, usually two to four inches ahead of the rear axle. The unseen "meat" of the hitch is often an integral part of the mounting rails beneath the bed. On other models, the hitch includes a steel plate bolted through the floor of the bed to the rails. The ball attaches through the center of the plate. Depending on the hitch model, the ball can be permanently mounted, removable, or reversed or folded down out of the way when not in use.

The trailer portion of a gooseneck hitch is rated from 20,000 to 30,000 pounds depending on trailer design and construction. On the trailer, the gooseneck coupler usually slides into a steel tube (called the tongue or neck) attached to the bottom of the bunk. If a gooseneck is not level when attached to a truck, you can either adjust the suspension to lower the truck or adjust the neck of the gooseneck coupling. Some gooseneck trailers have several inches of vertical adjustment in the neck, which allows you to raise or lower the trailer to fit trucks of various heights. To adjust the height, the bolts in the gooseneck are loosened and the tube is moved up or down as needed. Be aware, however, that if you shorten the neck of a gooseneck to fit a tall truck, the trailer might contact the tailgate or bed rails when turning. Many new trailers are designed specifically to fit the higher beds of modern trucks.

## FIFTH-WHEEL HITCH

You'll know a fifth-wheel hitch when you see one, because it takes up a good portion of the truck bed. It looks like a smaller version of the hitch used by big-rig semis, and is sometimes called a mini fifth-wheel. It is basically an elevated flat plate mounted on the truck into which a pin on the trailer fits.

A platform, often consisting of four legs, is bolted through the bed to the rails. The platform supports the head of the hitch. The head consists of the head plate, the jaws, and the handle. The head plate (also called the bearing plate or wheel) is a flat plate with a hole in the center. The plate supports the trailer and allows it to rotate. The jaws (some have one, others two) hold the kingpin securely in place. A handle locks and releases the jaws.

On the trailer, a pin box attaches to the bottom of the bunk and is the counterpart of the tube on a gooseneck coupler. The kingpin plate attaches at the bottom of the pin box and rides on the head plate of the fifth-wheel hitch. (Grease is commonly used between the two plates for lubrication, but a nylon lube plate, like a giant washer, is less messy). The kingpin, or pin, projects from the center of the kingpin plate and fits into the head plate hole where it is locked in place by the jaws, connecting the trailer to the truck.

## Hitch Load

A straight-pull hitch carries 10 to 15 percent of the gross trailer weight, whereas a gooseneck or fifth-wheel hitch carries 20 to 25 percent. A ¾-ton or larger truck is better than a ½-ton for towing gooseneck trailers because it has a larger payload capacity. Exactly how large a truck you need depends on the size of your trailer. If a loaded trailer weighs 16,000 pounds and 25 percent will be carried by the hitch, then you need a truck with enough payload capacity to carry 4000 pounds of tongue weight plus any cargo you would put in the truck.

# Types of Gooseneck and Fifth-wheel Hitches

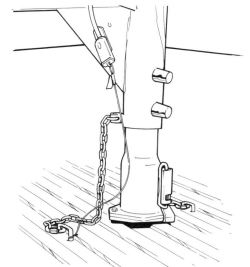

**Gooseneck hitch hitched**

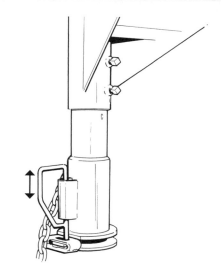

**Adjustable gooseneck tongue**

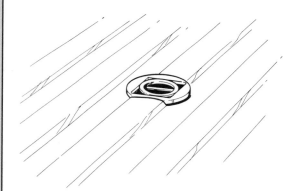

**Gooseneck ball, turnover or fold down, flush with bed**

**Gooseneck ball recessed in flatbed**

**Gooseneck ball in bed with eye bolts for safety chains**

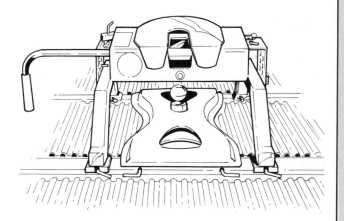

**Fifth-wheel/gooseneck combination in pickup bed**

## GOOSENECK HITCH VS. FIFTH-WHEEL HITCH

Whether you choose a gooseneck hitch or a fifth-wheel hitch depends largely on two factors: how much you value the space in the bed of the truck when towing and which hitch you are most comfortable connecting.

The main advantage of gooseneck hitches, and the reason most horse haulers prefer them, is that the ball takes up minimal space in the truck bed, so you can use most of the bed for cargo even with the trailer hooked up. Removable and fold-down balls enable you to utilize the entire bed when the trailer is not attached. A fifth-wheel hitch, on the other hand, takes up a considerable portion of the truck bed. Although you can purchase a removable type that frees up the entire bed when you are not using the trailer, much of the bed is unavailable when towing. A removable fifth-wheel hitch is also heavier and more difficult to lift in and out of the truck bed than a simple gooseneck ball.

The main reason people choose fifth-wheel hitches over goosenecks is ease of hookup, which is also why they are so popular on travel trailers. On level ground especially, there is no contest. The head plate of the fifth-wheel has a funnel channel that guides the trailer pin into the hole as you back up, making it a simple one-man operation. A wider kingpin funnel area makes the process even easier.

**Although a gooseneck hitch uses very little bed space, it still is important to keep cargo clear of the coupler to allow for proper movement when turning**

With a gooseneck hitch, you have to maneuver the truck until the ball is almost directly under the trailer coupler, which can be tricky.

On uneven terrain, however, a gooseneck might have a slight advantage over many fifth-wheel hitches because the tilt of the truck doesn't affect the gooseneck's ball and socket alignment as much as it does the double-plate alignment of a fifth-wheel. Newer fifth-wheel hitches tilt in all directions, however, so they can align more easily on uneven ground than less-flexible older models.

Another thing to consider is that fifth-wheel hitches do not require safety chains like gooseneck hitches do. Crawling into the bed to connect the chains on a gooseneck can be messy, inconvenient, and awkward. And finally, a gooseneck hitch costs less to buy and to install than a fifth-wheel hitch.

# Gooseneck or Fifth-wheel Trailers vs. Straight-pull Trailers

The gooseneck design is often used for trailers that are considered too heavy for a straight-pull hitch. The maximum tongue weight of a straight-pull hitch, for example, is around 1000 pounds; a fifth-wheel or gooseneck hitch can carry as much as the tow vehicle can handle. There are several other advantages and some disadvantages to consider when deciding between a gooseneck or fifth-wheel hitch and a straight-pull hitch.

## GOOSENECK AND FIFTH-WHEEL ADVANTAGES

Since the weight is distributed more evenly between both axles of the towing vehicle, the truck remains more stable with a fifth-wheel or gooseneck trailer, because the trailer is less susceptible to sway in wind and on curves and is less likely to fishtail (see chapter 11 for more on fishtailing). A straight-pull trailer, being attached behind the rear axle of the towing vehicle, has considerable leverage to push the rear

# Hooking Up a Gooseneck or Fifth-wheel

Gooseneck and fifth-wheel hitches can be easier to hook up than straight-pull hitches. The process goes more smoothly with a competent helper to watch the hitch and direct the tow vehicle if necessary. Make sure driver and director agree on a set of hand signals. The director should be in view of the driver at all times, and the driver should proceed slowly and cautiously. If you must hook up alone, be sure to set the vehicle's parking brake whenever you get out to check alignment of the hitch or to adjust the elevation of the trailer.

**1** Before you begin, make sure that the hitch is in the unlatched position. Place wheel chocks under both the front and back of the trailer's wheels on both sides. If the tow vehicle has a tailgate, lower it and back the vehicle to within a few feet of the trailer. Tailgates are a common casualty with goosenecks and some people just remove them altogether.

**2** **Gooseneck:** Adjust the trailer height so the coupler is one or two inches above the ball on the tow vehicle.
**Fifth-wheel:** Use the trailer jacks to adjust the height of trailer so the kingpin plate is level with the head plate on the tow vehicle.

**3** **Gooseneck:** Slowly back the vehicle until the ball is directly under the trailer's coupler. Lower the jacks until the coupler is fully engaged on the ball.
**Fifth-wheel:** Slowly back the tow vehicle until the kingpin is completely inserted in the hitch.

**4** Close the latch. Some fifth-wheel hitches lock automatically when the pin engages the jaws, and others need to be locked manually by moving a lever after the pin is in place. Some hitches have a kingpin indicator that tells you if the hitch is locked in place.

**5** Insert the lock pin in the latching lever.

**6** Plug in the light/brake cable to the trailer. With a gooseneck, hook up the safety chains.

**7** Raise the trailer jacks to their fully retracted position.

**8** Close the tailgate and remove the wheel chocks.

**Newer trucks have higher beds that don't allow old gooseneck trailers to ride level without modifications.**

**A common result of turning too sharply with a straight-pull trailer.**

end of the vehicle around (which is bad), whereas a fifth-wheel or gooseneck, because it is over the rear axle, has little leverage (which is good).

Fifth-wheel and gooseneck trailers are more maneuverable, especially in tight spots, because the truck and trailer can be turned at 90 degrees to one another. When backing a straight-pull trailer or turning very shortly, you have the potential to collide the rear corner of the truck with the trailer.

Fifth-wheel and gooseneck trailers are easier to hook up because the driver generally can see both the hitch and the coupler through the back window of the cab. Many ingenious devices have been invented to make hooking up a straight-pull trailer easier, but especially when hooking up alone, you might need to get out of the vehicle and check the alignment several times.

Fifth-wheel and gooseneck trailers have more ground clearance when going over bumps and dips. A straight-pull hitch is more likely to hit the ground because the front of the trailer often has only a few inches of clearance.

The bunk (nose area) over the gooseneck provides additional storage or sleeping space.

## Jack Notes

A gooseneck trailer has two manual or electric jacks near the front, which raise and lower the front of the trailer. After the trailer is hooked up and before you head out, raise the jacks as far as they will go. Some jacks have a safety mechanism that clicks when they reach their upper limit; quit raising the jack when it clicks.

Remove the lock pin in each jack, slide the pad post up into the tube, and lock it in the highest position. This will give you maximum clearance between the ground and the bottom of the jacks and minimize the risk of damaging the jacks.

When hitching a fifth-wheel, double-check to be sure the hitch is securely engaged and latched before raising the jacks.

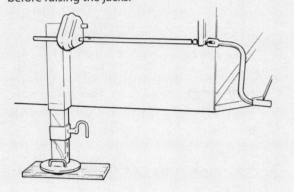

## STRAIGHT-PULL ADVANTAGES

A straight-pull trailer can be used with any vehicle that is rated for towing, including an SUV, van, or pickup with a topper or tonneau cover. They are generally universally adaptable between vehicles as long as the vehicle is equipped with a receiver, the proper size ball, and a plug for brakes and lights. A straight-pull trailer is easier to borrow, loan, and share, and makes it likelier that you will find an emergency tow if your truck breaks down.

You can easily hook a straight-pull trailer to a tractor or other vehicle with a ball in order to park it or move it around your property. A fifth-wheel or gooseneck trailer is generally restricted to use with an open-bed truck with a compatible hitch.

With a straight-pull trailer, you have full use of the tow vehicle bed whether or not the trailer is hooked up.

In some ways, it is easier to hitch a straight-pull trailer than a gooseneck. With a gooseneck, you usually have to climb into the truck's bed.

Since less tongue weight is transferred to the truck from a straight-pull trailer, if all other things are equal, it is possible to use a truck with less towing capacity.

You have better visibility out of the rear window of the tow vehicle with a straight-pull trailer because the trailer is farther away. The bunk of a fifth-wheel or gooseneck trailer obscures a good portion of your view. A straight-pull trailer also tracks closer to your tow vehicle than a gooseneck, so you don't have to take corners as wide.

A straight-pull trailer costs about 20 percent less than a similar fifth-wheel or gooseneck trailer.

Straight-pull trailers are typically shorter so they require a smaller storage space.

Although hitch selection is of top importance when choosing a trailer, there are many more features to consider, as outlined in the next chapter.

## The Long and Short of It

A long-bed pickup is better suited than a short-bed pickup for towing a gooseneck trailer that has a bunk. The longer wheelbase gives more stability, and the hitch is located farther from the cab so there is more clearance when turning sharply. Many a short-bed owner has shattered the rear cab window or dented the truck when making short turns.

There are gooseneck and fifth-wheel slider hitches available especially for use with short-bed trucks. This type of hitch slides back and forward manually or automatically into "maneuver position" to allow more clearance for safe backing and turning sharply at slow speed without hitting the cab. Another short-bed solution is a gooseneck extender coupler that slides up into a standard round trailer tube to replace the standard coupler and increases the distance between the cab and the trailer bunk.

**This flatbed gooseneck has no bunk so it fits well with a short-bed truck.**

**A long bed truck provides ample clearance between cab and bunk.**

# 9 Trailer Features

Whether your budget dictates buying a new or used trailer, you must make some basic decisions. There are many factors to consider when investing in a horse trailer. Small but important details can make a trailer a dusty, rough-riding oven or a comfortable, safe coach. Learn all you can about trailers and then make your personal wish list, giving your horse's safety and comfort top priority.

# Types of Trailers

There are three types of horse-hauling vehicles: the enclosed trailer, the stock trailer, and the horse van. Each has advantages and disadvantages, depending on your needs.

## ENCLOSED TRAILERS

Enclosed trailers are the most common trailers seen on the road and come in one-, two-, three-, four- and six-horse models. Horses travel standing in tie stalls. Enclosed trailers can range from simple models with under-manger tack compartments to luxurious trailers with large tack rooms, feed rooms, and living quarters for people.

## STOCK TRAILERS

Stock trailers are usually the equivalent of a four-horse trailer in length and style, but the sides are slatted and the interior is open, so they weigh less than a similarly sized enclosed trailer. Stock trailers are commonly available in 16- to 32-foot lengths. They are designed to haul horses either loose or tied in box stall-sized spaces. If you are buying a stock trailer, take care to select one that is tall enough for horses. Many short stock trailers that are perfectly suitable for cattle, sheep, or hogs do not have enough clearance for horses. (See recommended dimensions on page 117.)

## A Horse's Trailer Wish List

- ❏ A light, airy space
- ❏ Lots of headroom
- ❏ Room to spread out legs from side to side
- ❏ Ventilation and clean air
- ❏ Room to stretch neck and head forward and to lower head to cough and blow
- ❏ Comfortable, non-slip footing
- ❏ No sharp edges or protrusions

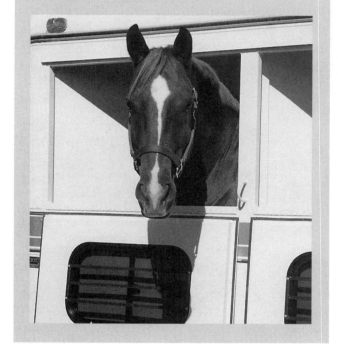

**Enclosed trailer**

**Stock trailer**

## VANS

A van is essentially a stable on wheels, with the truck and horse compartment combined. They can be 13 to 26 feet long and can carry up to nine horses. Vans generally provide a more comfortable ride for the horse, because they have better suspension, which is less fatiguing to a horse's legs. They tend to be heavier and better insulated, which results in less road vibration and noise and contributes to a more stable temperature. Vans typically allow more head room for the horses, and the stalls often face each other so that horses can look at each other. Vans are the most expensive of the three types, ranging from $40,000 to more than $100,000.

**Van exterior**

**Van interior**

# Floor Plans

Trailers come in an almost limitless variety of floor plans, especially when you consider that many trailer manufacturers will build a horse trailer to your custom specifications. You have many choices when it comes to the number of stalls, the configuration of stalls, the size and shape of the tack room, dressing room, and even living quarters in a horse trailer. Each has its advantages and drawbacks. Some common configurations are shown in the figures on pages 112–113.

## OPEN-FLOOR TRAILERS

Open-floor trailers (often synonymous with stock trailers or any trailer that has removable dividers) have large open spaces and are quite versatile. They can be fully enclosed with solid walls or have slatted sides like a stock trailer. Horses can be tied in an open trailer or they can be hauled loose. There is usually a movable center divider that separates the front half from the back half.

Some horses exhibit less stress when they ride in a stock trailer with slatted sides (compared to a fully enclosed trailer) because they can see out and can move around and find a comfortable traveling stance. This comes in handy when transporting mares and foals and young horses. However, in trailers with slatted sides, horses can get dusty, cold, and/or wet.

Limited research has shown that given a choice, most horses like to ride facing the back of the trailer at an angle. But individual horses have preferences. However, if more than one horse is being hauled in a stock trailer, whether they are loose or tied near each other, they must get along very well, or they could injure each other. Horses moving about can contribute to trailer sway.

With open-floor trailers, you need to use hay bags on the walls for feeding.

Mats should fit snugly to stay in place or be fastened down. Thin, narrow strip mats can easily become dislodged by scrambling horses and get tangled up in their legs.

**Advantages**

- Open and inviting
- Comparatively lightweight
- Less expensive than enclosed trailers
- Horse can find most comfortable way to stand
- Horse can be turned around and unloaded walking forward
- Versatile; useful for hauling hay, appliances, and other large items

**Disadvantages**

- Horse can get dusty, wet, or chilly from open sides (sides on some can be closed with non-sliding Plexiglas)
- Horses can fight, be injured, or cause too much movement for safe trailering

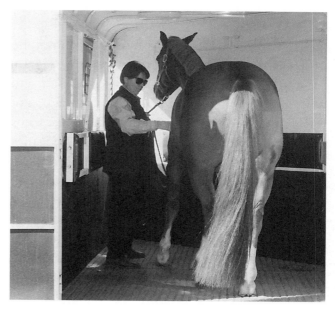

**Open-floor trailer**

## SIDE-BY-SIDE TRAILERS

Straight-load trailers, also called side-by-side, have been the most common two-horse trailers for many years. Two-horse steel trailers usually are the least expensive fully enclosed trailer you can purchase. They also come in four- and six-horse models. Some four- and six-horse side-by-side trailers are configured so that some of the horses face forward and some face rearward.

Straight-load trailers are considered fairly safe in the event of an accident, because if the trailer overturns, the horses lie flat, and sometimes the rig can be righted without unloading them. If it is possible and necessary to remove the horses with the rig on its side, the top horse can be slid out on the center divider, then the divider can be removed to release the bottom horse.

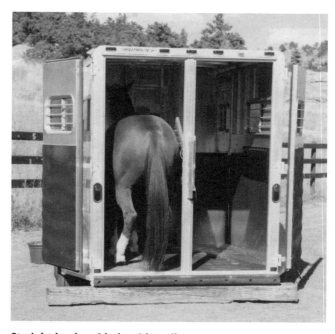

**Straight-load or side-by-side trailer**

**Advantages**

- Most common type with largest selection in used market
- Often lower priced than slant-load
- Often has a manger to feed horses hay and grain
- Easy access to horses
- Same width as towing vehicle, so tires follow truck tracks

**Disadvantages**

- Horses may feel restricted by manger wall at forelegs
- Horses may scramble against sidewalls to maintain balance
- Horse must do majority of bracing on front legs so this configuration is hard on horses with navicular or laminitis

## SLANT-LOAD TRAILERS

Slant-load trailers are a relatively recent innovation. In a slant-load, the horses ride diagonally. Horses seem to be more comfortable with acceleration and deceleration if they stand at an angle to the line of travel so that their diagonal pairs of legs take turns resisting the various forces of acceleration and stopping. But as the right front and the left hind take most of the stress, some experts feel this is not good, especially for horses that spend a lot of time on the road. The slant-load design allows more horses to be hauled in a shorter but slightly wider trailer. Federal law limits trailer width to 8½ feet, so stall length is limited and slant-load stalls aren't generally long enough for many horses larger than 16 hands.

There are safety concerns with a slant-load; if this kind of trailer overturns, it can be difficult to get the horses out without righting the trailer or cutting the top off.

The odd triangle of space that is left over at the back of a slant-load is often wasted. To utilize this space, some trailers make it a permanent tack room, but this restricts the wide loading opening, which is one of a slant load's most inviting features to a horse. With the opening as narrow as that of a straight-load trailer, you either have to enter ahead of your horse or send the horse in and then follow to tie him, neither of which is a good option. A more workable option for the space is a collapsible, removable, or swing-out saddle rack/tack room that moves, with varying levels of effort, to allow a wide doorway for loading and unloading horses.

**Advantages**

- Shorter trailer (one to four feet shorter, depending on trailer)
- Open and inviting
- Many have option of turning horse around to unload
- Can be useful for hauling bulky items such as hay and equipment

**Slant-load trailer**

**Disadvantages**

◆ Slightly wider trailer (one to two feet wider) so tracks are different from truck and trailer tire

◆ Can hit curb or drop off edge of road if not careful

◆ More expensive than straight-pull or stock

◆ Usually best suited for small to medium-sized horses

## IN-LINE TRAILERS

In-line trailers are no longer in production, but used ones are available in one- and two-horse models. In a two-horse in-line, the horses travel facing forward in single stalls, one in front of the other. The second horse is loaded behind the first horse, head to tail, and does not have a manger.

Because the axles of an in-line trailer are spread far apart, like a wagon, this is a very stable trailer with a very light tongue weight. With its narrow profile, it stays on the road better than a side-by-side trailer, and is thought to be one of the safest designs in the case of an accident because the horses don't get scrambled together and can be removed either through the roof or the back door.

In-line trailers can be very difficult to back up, however, as the trailer pivots at the hitch as well as at the front axle, making it maneuver more like a hay wagon. It is somewhat like pushing a rope down the street.

**Advantages**

◆ Easy to pull

◆ Safer in an accident

**Disadvantages**

◆ Not readily available

◆ Hard to back up

## REVERSE-LOAD TRAILERS

Reverse-load trailers have both front and rear entries, allowing the horses to load through one door and exit through the other without having to turn around or back up. Reverse-load trailers are

**In-line trailer**

specially engineered and balanced for the horses to ride facing either forward or backward.

*Never* load your horse facing rearward in a trailer that isn't specifically designed for it. If you loaded your horse facing the rear in a conventional two-horse trailer, the horse's weight would be too much to the rear, causing trailer sway.

**Advantages**

◆ Horse can load and unload without backing up

◆ Horses can ride facing forward or backward

**Disadvantages**

◆ Often a custom trailer, so expensive

# Loading Entry

Horses load by stepping up into the trailer or by walking up a ramp. There are advantages and disadvantages to either style, but it is a good idea to train yourself and your horse to be comfortable loading both ways.

## STEP-UP

The step-up style is simplest, least expensive, and the most common entry style. The open doors help guide a horse when loading. When loading on a high step-up, a horse could bang a shin on the floor sill; when unloading, a hind leg could slip and get caught

# Floor Plans for Various Trailers

manger (top door) and tack compartment (lower door)

manger

escape doors

center divider

**Straight-pull two-horse in-line with one rear door, one manger door, and one tack compartment in front, four escape doors, and four sliding windows.**

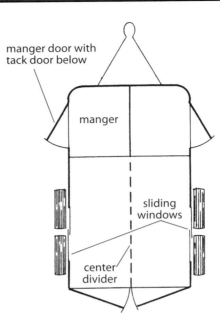

manger door with tack door below

manger

sliding windows

center divider

**Straight-pull two-horse straight-load with two rear doors, two manger doors, two under-manger tack compartments, and two sliding windows.**

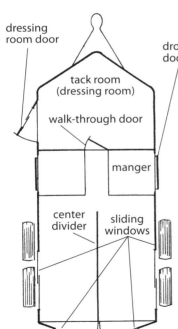

dressing room door

drop-down door

tack room (dressing room)

walk-through door

manger

center divider

sliding windows

**Straight-pull two-horse straight-load with walk-through to tack room, two rear doors, two drop-down manger doors, four sliding windows, and a tack room door.**

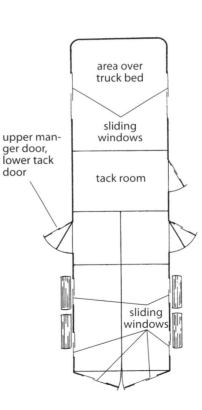

area over truck bed

sliding windows

upper manger door, lower tack door

tack room

sliding windows

**Gooseneck two-horse straight-load with tack room, two rear doors, two manger doors, two tack compartments, tack room door, and eight sliding windows.**

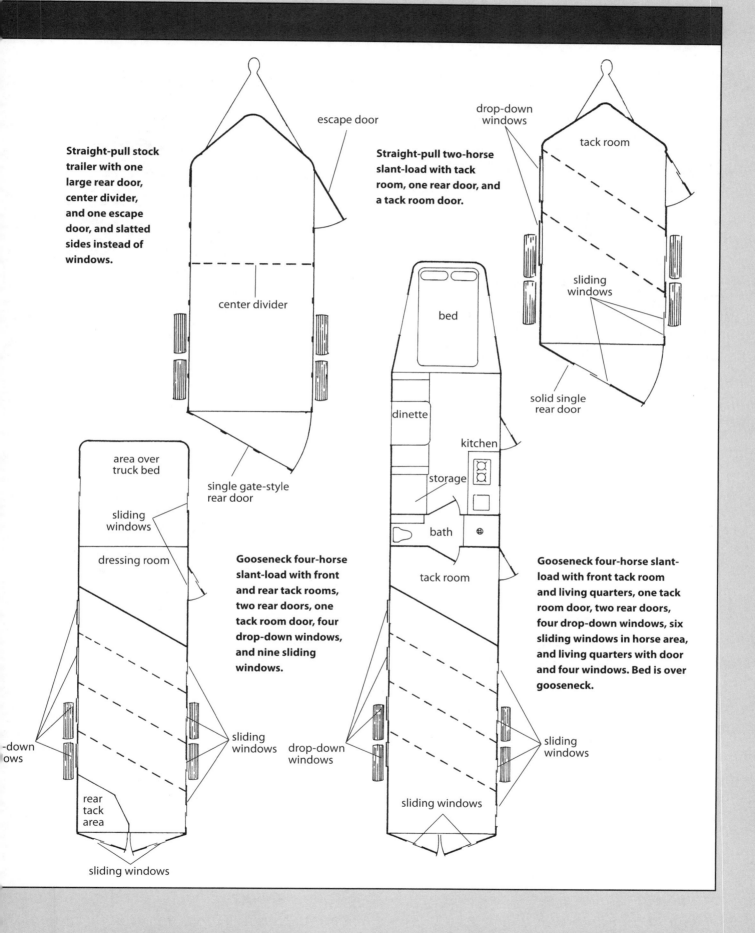

**Straight-pull stock trailer with one large rear door, center divider, and one escape door, and slatted sides instead of windows.**

escape door

center divider

single gate-style rear door

area over truck bed

sliding windows

dressing room

**Gooseneck four-horse slant-load with front and rear tack rooms, two rear doors, one tack room door, four drop-down windows, and nine sliding windows.**

-down ows

sliding windows

rear tack area

sliding windows

drop-down windows

drop-down windows

**Straight-pull two-horse slant-load with tack room, one rear door, and a tack room door.**

tack room

sliding windows

solid single rear door

bed

dinette

kitchen

storage

bath

tack room

**Gooseneck four-horse slant-load with front tack room and living quarters, one tack room door, two rear doors, four drop-down windows, six sliding windows in horse area, and living quarters with door and four windows. Bed is over gooseneck.**

sliding windows

sliding windows

Reverse-load trailers have both a front and a rear entry. The horses load in one door and go out the other without having to turn around or back up. A horse can travel facing forward or rearward.

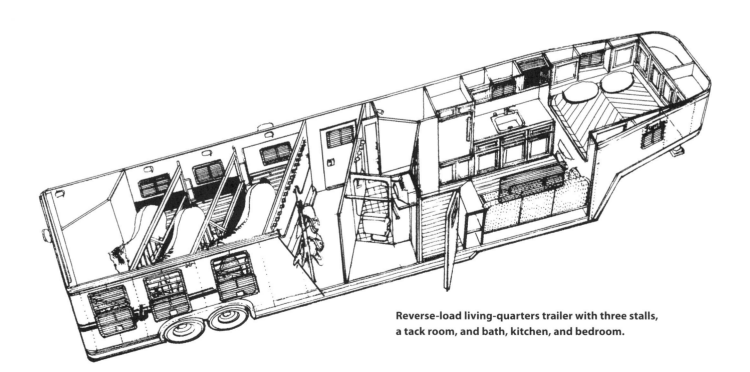

Reverse-load living-quarters trailer with three stalls, a tack room, and bath, kitchen, and bedroom.

underneath. For safety, choose a trailer with a 16-inch step-up height or less. Twelve inches is common and works well. All step-up trailers should have a rubber bumper or round steel bumper running the entire width of the rear of the trailer. Take care when unloading horses in wet weather or on slippery ground or grass. That's when a horse is most likely to slip under the trailer sill.

## RAMP

A good ramp can be safer than a step-up entry. A well-constructed one-piece ramp makes loading young horses in high trailers easy and helps a lame or old horse to load more comfortably. Some step-up trailers are difficult for a short horse to negotiate, especially if the step is more than 16 inches off the ground. The gradual incline of a ramp 48 inches or longer makes it easy for a small horse to load. Ramps also allow you to easily load a garden tractor or an appliance on a dolly.

Not all ramps are created equal and ramps in general can have drawbacks. A ramp costs more and adds substantial weight (an extra 200–300 pounds) to the rear of a trailer where you don't want it. Some horses will spook or balk at the sight of a ramp; others will startle when they hear the hollow sound or feel the ramp move when they step on it. It's more important to have your trailer parked on level ground when loading with a ramp, to give it stability.

With a ramp, there are often no lower trailer doors to guide the horse when loading. A horse could duck off to the side and scrape his back under the top door or simply step or fall off the ramp when loading. The hinge between the ramp and the floor catches debris, which must be cleaned out before closing the ramp; the crevice can be a potential danger for a small hoof.

The length of the ramp and the height of the trailer floor dictate the slope the horse must climb. A too-short ramp makes for a steep climb. One-piece ramps can be very heavy and hard to lift up. A power-assist or a well-balanced spring device makes it easier for you to close the ramp, but you still must

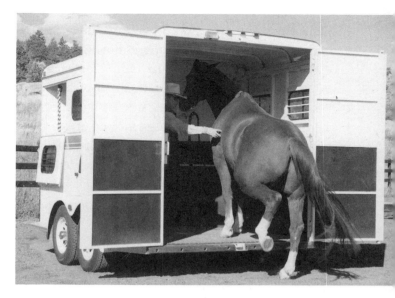

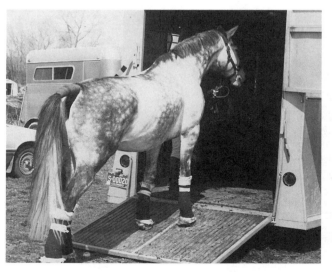

**Top: Step-up open-floor; Middle: Step-down straight-load;
Bottom: Wide ramp**

bend down and pick it up. The ramp's hinge and the spring or power-assist add more moving parts that can require repair. Individual ramps for each side of the trailer are lighter and easier to pick up, but because they are narrower, it is easier for a horse to step off or slip off. Side ramps located at the front of a larger trailer allow you to walk the front horses forward to unload or to easily load horses in trailers where they ride facing backward. Ramps are especially appropriate for four- and six-horse models so you can unload some of the horses through the front, and back the rear horses out the rear.

# Trailer Materials

Trailers are most commonly constructed of steel, aluminum, fiberglass-reinforced plastic, and proprietary synthetics. An all-steel trailer is strong, but it is also prone to rust and is heavier than its aluminum counterpart. Aluminum trailers are lighter, but are more easily damaged on impact. Fiberglass is used primarily for roofs and fenders, where its light weight is a desirable attribute. A trailer with a steel frame, aluminum skin, and fiberglass fenders seems to blend the best of all worlds.

## STEEL

The short, two-horse steel trailer with an under-manger tack compartment was the most common horse trailer configuration for many years and still is a popular, economic choice. Often such a trailer is a straight-load, step-up trailer with an escape door, two manger doors, and one or two doors to the under-manger tack storage.

A trailer with a steel frame and skin is sturdy but heavy, and is subject to rust, which requires repair and painting. The lower inside walls, if not reinforced when new, eventually must be reinforced or replaced because of damage from hooves and shoes.

## ALUMINUM

All-aluminum trailers weigh the least, cost the most, and last the longest. More than half of new trailers sold are aluminum. Trailers with an aluminum frame and skin cost from 30 to 50 percent more than equivalent steel models, but they are 20–30 percent lighter. They require very little maintenance and hold their resale value better than other trailers. Aluminum doesn't need repainting, but it dents more easily than steel, and damage to an aluminum trailer is likely to be greater in the event of an accident.

**Side ramp — note unlevel ground**

**Narrow rear ramp**

**Steel trailer**

**Aluminum trailer**

### STEEL FRAME, ALUMINUM SKIN

Trailers with a steel frame and aluminum skin weigh less than all-steel trailers, don't need repainting, and are less prone to rust, but they are almost double the price of a steel trailer. Aluminum diamond plate is often installed on trailer fronts and fenders to protect the trailer from gravel damage caused by the trucks' rear tires. Because galvanic corrosion can attack aluminum when it is fastened to steel, trailers with a steel frame and an aluminum skin require special taping and insulation in all places where the two metals contact each other.

### FIBERGLASS

Fiberglass-reinforced plywood trailers make up a small percentage of the used-trailer market. They are built on a steel or aluminum frame with a skin of fiberglass-covered plywood. The resulting trailer is the weight and price of a steel one, but may require less maintenance.

Fiberglass is often used for roofs and fenders of various trailer models, as it is cool, lightweight, and easy to repair.

## Trailer Specs

When choosing a trailer, take into consideration the size of the horses you plan to haul and their traveling habits.

### HEIGHT

The inside height of trailers ranges from 72 inches to 90 inches, though most modern ones measure around 84 inches. The height inside an old-style horse trailer is 72–78 inches. Newer standard-sized trailers range from 80 to 86 inches, but it is not uncommon to find warmblood and draft-horse trailers as tall as 96 inches inside. Be aware that the height at the entry might be four inches lower than the interior ceiling, and the roof's support ribs are generally somewhere between the two. Whether you are shopping for a new or used trailer, always take a tape measure with you.

A 16.2-hand horse stands 66 inches at the withers. Loaded into a 78-inch trailer, he would only be able to raise his head 12 inches above his withers while loading or unloading and would have to carry it that low through the entire trip. A horse that size would be much more comfortable in a trailer with an 84- to 90-inch ceiling.

A 15- to 16-hand horse can fit into trailers with ceilings lower than 84 inches, provided he is level-headed about loading and unloading, but an 84-inch or taller trailer would be best. If you are planning to haul a very large horse, you may need to look into a custom trailer or van.

Measure a prospective trailer to see if it will fit in your storage building. Keep in mind the outside height of your trailer, including open vents and roof rack, so that when you pull under a marked over-hang, such as at a motel, you don't rip the roof off.

## LENGTH

In a straight-load trailer, length usually refers to the stall length (the distance from the butt bar to the chest bar) and corresponds to the horse's length from chest to rump. Trailer lengths range from 66 inches to 88 inches, with the average around 70. The average 14.2- to 15.2-hand horse measures about 66–68 inches from chest to tail, so a stall 72–78 inches long (from manger to butt bar) works well for many horses.

In a slant-load trailer, stall length corresponds to the length of the horse from nose to tail. The stall length in slant-load trailers ranges from 90 to 96 inches. The length of a slant-load stall is measured down the middle of the stall from one wall to the other to correspond to the horse's nose to tail measurement. A 15- to 16-hand horse is about 92–94 inches from nose to tail when his head and neck are in a natural, relaxed position (not compressed or stretching). Horses over 16 hands are at least 96 inches long from nose to tail. Since few slant load stalls are longer than 96 inches, they generally are not suitable for horses larger than 16 hands.

For stock trailers, allow eight feet of interior length for every two horses, so a 16-foot trailer can hold four horses, a 24-foot trailer can hold six horses, and so on.

## WIDTH

The width of a single trailer stall can range from 26 to 38 inches, with most measuring toward the lower end. The stalls in slant-load trailers are generally narrower than those in straight-loads. Warmblood trailer stalls can be as wide as 48 inches. The average 15- to 16-hand saddle horse does well in a standard-width stall as long as the stall divider does not go all the way to the floor. Some horses travel better in a narrower stall that offers some support than they do in a wide stall where they can move from side to side excessively.

## SAMPLE CURB WEIGHTS OF VARIOUS TRAILERS

| TRAILER TYPE | WEIGHT IN LBS (KG) |
| --- | --- |
| One-horse Euro-style, straight-pull trailer | 1300 |
| Two-horse aluminum straight-pull, straight-load with dressing room | 2500 |
| Two-horse aluminum straight-pull, slant-load, with dressing room | 2800 |
| Two-horse steel straight-pull, straight-load, tack compartments | 3000 |
| Two-horse steel straight-pull, straight-load with dressing room | 3300 |
| Two-horse steel straight-pull, slant-load with dressing room | 3800 |
| Two-horse steel frame aluminum gooseneck, slant-load with dressing room | 4100 |
| Four-horse aluminum gooseneck, straight-load with dressing room | 4500 |
| Four-horse aluminum gooseneck slant-load with dressing room | 4750 |
| Four-horse steel gooseneck straight-load with dressing room | 7000 |
| Four-horse steel gooseneck slant-load with dressing room | 7500 |

# Trailer Weight

It is critical to your horse's safety and yours to know the specific weights related to any trailers you are considering purchasing. Master the terminology and know what questions to ask. Although many salespeople are knowledgeable and helpful, this is one area where they can be less informed, so make it your responsibility to know all about weight.

## CURB WEIGHT

Your trailer dealer can supply the weight of an empty trailer without any add-ons or options. The actual curb weight will include the weight of any nonstandard add-ons such as 300 pounds for a rear ramp,

## Curb Weight Considerations

**FACTORS THAT AFFECT CURB WEIGHT**

**Style:** All other things being equal, a stock trailer will be lighter than a slant-load, which will be lighter than a straight-load.

**Skin:** Aluminum is lighter than steel or plywood, which are about equal in weight.

**Frame:** Aluminum is lighter than steel.

**Roof:** Fiberglass is lighter than aluminum, which is lighter than steel.

**ITEMS THAT ADD TO CURB WEIGHT**

**Floor mats:** Estimate 70–160 pounds for a one-horse mat, depending on size and thickness.

**Sidewall mats:** Sidewall mats for a two-horse trailer can add as much as 200 pounds to the curb weight.

**Ramp:** Will likely add 200–300 pounds.

**Spare tire:** A spare tire will add about 40–60 pounds.

**Accessories in the tack room:** Carpeting, saddle racks, tack trunk, brush bins, or other accessories add considerable weight.

200 pounds for sidewall mats, and so on. The best way to determine the actual curb weight of your trailer is to weigh it empty on a commercial scale. Be sure the tack room is empty. Stop with just the trailer wheels and tongue jack on the scale, block the trailer's wheels, and detach the trailer coupler from the truck's receiver. If you don't detach the trailer, part of its weight (tongue weight) is being borne by the truck and you won't get an accurate measurement.

### GROSS VEHICLE WEIGHT

The gross vehicle weight (GVW), or loaded trailer weight (LTW), refers to the weight of the trailer and all of its cargo. The most accurate way to determine your trailer's actual GVW is to weigh it on a scale when it is fully loaded. To estimate your trailer's GVW, add the manufacturer's curb weight, the weight of any add-ons, the weight of the horses, and 100–200 pounds per horse for feed, water, and tack. A thousand pounds is often used as a standard weight per horse, but you should know each of your horse's weight (via weight tape) for medical purposes, so you can use their actual weights. Your trailer's GVW should never exceed the trailer manufacturer's gross vehicle weight rating (GVWR) for the trailer. If you overload your trailer, the trailer's suspension or brakes could fail. Your trailer's GVW should never be greater than your truck's maximum tow weight

rating (MTWR). If you try to tow a trailer that has a greater GVW than the truck is rated to tow, it would not only be dangerous to drive, but also could result in damage to the truck's engine, drive train, suspension, brakes, and tires.

## Trailer Brakes

A separate trailer brake system with a manual override lever within the reach of the driver should be an integral part of any horse trailer rig. State laws dictate which trailers require brakes, whether brakes are required on all four wheels, and if a breakaway control is required. All states require brakes on any trailer with a GVW of more than 3000 pounds, which includes all horse trailers. In most cases, brakes are required on all axles. A four-wheel brake system is safer and more efficient than a two-wheel brake system, even for a two-horse trailer, and brakes on all wheels are imperative for larger rigs.

Electric brakes are the most common. They connect to your truck's electrical system and are activated when you step on the brake pedal, use the hand controller, or when the emergency trailer breakaway system is activated. Some manufacturers offer vacuum-hydraulic brakes, which are independent of the tow vehicle and are activated when pressure is exerted on the hitch coupler, as when you are

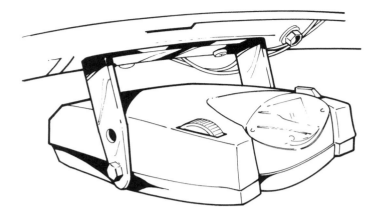

**Electric brake controller under dashboard**

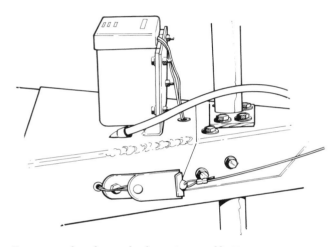

**Emergency breakaway brake system and battery**

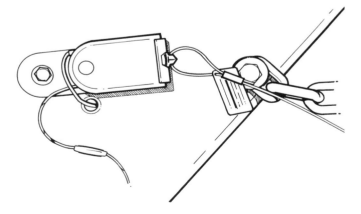

**Never leave the pin out of the breakaway switch any longer than necessary, because while it is out, electricity is powering the brake magnets. In a matter of minutes, the battery will be completely drained.**

slowing down. The system is controlled by a hydraulic cylinder located at the front of the trailer. Some Euro-style trailers use a mechanical brake system similar to the coupler-activated hydraulic system.

### EMERGENCY BREAKAWAY SYSTEM

Federal traffic safety law mandates that all trailers that require brakes have an emergency breakaway braking system. The breakaway brake mechanism is a battery-operated safety switch that activates your trailer brakes if your trailer should happen to come unhitched from your truck. The system is powered by a separate battery that is usually located in the tack room (the best option) or on the tongue of the trailer. The switch that activates the brakes is usually located on the tongue of your trailer and has a breakaway pin with a cable attached.

When you are towing, this cable should be securely attached to the frame of your truck. If the trailer were to separate from the truck, the pin would pull out of the switch, activating the trailer brakes. In order to work properly, the breakaway cable should be slightly longer than the safety chains. That way, if the trailer hitch breaks or becomes disconnected but the chains still holds, you can use the manual trailer brake controller to gradually apply the trailer brakes, rather than having the brakes lock up suddenly and possibly snap the safety chains.

# Suspension

Good quality trailer suspension should be sturdy but not stiff. Horse trailers usually either have leaf springs (old-style trailers) or rubber torsion bars (newer trailers). Rubber torsion suspension, which incorporates large sections of dense rubber to minimize bumps, vibration, and noise, is said to give a better ride than leaf-spring suspension, at least on highways. Leaf springs deliver a stiffer and somewhat noisier and rougher ride, but they absorb the drop from a deep hole or a rut better than rubber torsion, so they are better for ranch roads or travel in the backcountry or across pastures.

# Roof and Sidewalls

It is best if the trailer roof is one piece. Two- and three-piece roofs invite leaking and often require re-caulking. A square roof provides a full height across the entire width of the trailer in contrast to a rounded roof, which can have as much as a six-inch difference between the peak and edges of the roof. Many trailer roofs are a one-piece molded structure with a steel framework that acts like a rollover protective structure on a tractor.

The sidewalls should be thick and made of material that will resist the destructive forces of a horse's hooves. They can be single, double, or triple layers. Double and triple walls add more weight to the trailer but they provide insulation, making the trailer cooler in the summer and warmer in the winter. Insulated walls are sturdier and they decrease road noise.

# Flooring

Most steel trailers have wood floors. Either oak or pine is fine as long as they are No. 1 quality boards. Usually 2×6 or 2×8 boards are used, and they should be of a consistent thickness, straight-grained, and without knots or warps. Pressure-treated wood resists rot from manure and urine longer than untreated wood. Sometimes pressure-treated wood floors are painted for additional protection and ease of cleaning.

Most aluminum trailers have aluminum plank floors, which can be noisier and hotter than wood floors, but are lighter and won't rot. Aluminum can corrode from urine if not kept clean and dry; floors often have a drain hole at the rear for urine drainage.

## FLOOR MATS

The floorboards should be covered with removable rubber mats at least ⅝-inch thick to provide cushion and traction for the horses, and to protect the floorboards. Half-inch mats, which are standard in some trailers, are easier to remove but don't last as long or provide as much cushion as ⅝-inch mats. Three-quarter to one-inch mats offer more cushion but are very heavy to remove for cleaning.

## Tack Room Options
(check with your dealer)

- ❏ Trunk/bin/locker
- ❏ Water tank
- ❏ Saddle racks
- ❏ Bridle and tack hooks
- ❏ Cabinet with mirror
- ❏ Wardrobe bar and hat rack
- ❏ Blanket bars
- ❏ Brush bins
- ❏ Door organizer
- ❏ Lights
- ❏ Fan

# Tack Room/Dressing Room

A tack room can range from a small compartment under the manger with just enough room for a saddle and bridle to an extensive walk-in dressing room. A tack room/dressing room will add about four feet to the length of a trailer. The doors should be at least 30 inches wide and should lock securely. A door on the driver's side is most convenient for quickly grabbing a halter. Also when you are saddling your horse, you can carry the saddle directly from the tack room to your horse's near side. All items in the tack room should be able to be securely stowed and all doors should be lockable. That way in the event of an accident or sudden stop, the doors won't fly open and allow items to enter the stall area or fall onto the roadway.

# Doors and Windows

Doors and latches should work easily. Slam latches are the handiest and flush handles are the safest. Check that the doors and latches are strong and that they fit squarely, or they might be difficult to close or latch.

Escape doors (also called emergency or walk-through) come in many styles and sizes, but they

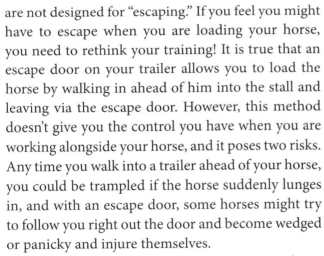

**A front dressing room**

**A rear tack room**

are not designed for "escaping." If you feel you might have to escape when you are loading your horse, you need to rethink your training! It is true that an escape door on your trailer allows you to load the horse by walking in ahead of him into the stall and leaving via the escape door. However, this method doesn't give you the control you have when you are working alongside your horse, and it poses two risks. Any time you walk into a trailer ahead of your horse, you could be trampled if the horse suddenly lunges in, and with an escape door, some horses might try to follow you right out the door and become wedged or panicky and injure themselves.

Instead, think of escape doors as access doors that allow you to check on your horse from the side or front. Some full-size access doors on the front of slant-loads have a hinged bar across the middle of the door to discourage the horse from coming out. Some are designed to unload horses and some are not. If you have a walk-through door to your tack room, you can lead the horse into the trailer by going

into the stall ahead of the horse and leave via the tack room. Even so, you risk being trampled if the horse rushes.

## REAR DOORS AND CENTER POST

Rear-door configurations can include two double doors (one on the top and one on the bottom of each side, making four doors), two single doors, one wide door, or a ramp with a bottom door and two top doors.

It is tempting to leave the top half of a double door off or open to provide extra ventilation when traveling on extremely hot days, but this is generally not a good idea. Your horse will probably be filthy and stressed out from the debris and traffic noise that are sucked into the trailer like a vacuum. Cigarette butts thrown from vehicles could cause a fire in the bedding or hay. Rear doors with windows and screens are a better option.

Double rear doors with a center post are the most common for a straight-load trailer. The post

The permanent center post of this trailer makes loading crowded. (Don't try this at home!)

An open rear without center post gives more room for loading and unloading.

holds the stall divider and adds strength and stability to the trailer, but it makes the trailer less versatile than a fully open rear entrance. Some horses may be reluctant to load if they feel confined by the post or may bang into it when loading or unloading.

Double rear doors without a center post are common on slant-load trailers. The wide opening of the entrance is inviting to horses and gives you plenty of room to work, turn the horse around, and unload facing forward. In addition, the open doorway makes it easy to load hay or other large bulky items.

Some trailers have double rear doors with a ramp folded on the outside. When a slant load has a permanent tack room at the rear, the resulting space makes a narrow loading entrance.

## DOOR LATCHES

A protruding handle latch can injure a horse or damage tack, clothing, or the trailer itself. Slam latches that are flush with the outside surface of the trailer are safer in terms of snagging horse or human. Poorly

designed rear-door latches have been responsible for horses falling out of moving trailers. A typical rear-door bar latch is strong and can be locked. It often has rubber-covered handles and locks into pockets at the top and bottom of the trailer door frame.

All doors should have holdback fasteners to keep the wind from banging the door shut just as you are loading a horse or walking out of the tack room with an armload of tack. The rubber dimple and nipple-type door fastener is popular on newer trailers and smoother than the older hook and pocket type, but some are not as secure in high winds.

## WINDOWS AND VENTS

Horse trailers can become very hot, so it is important that windows and vents be well designed and well placed. All safety glass or Plexiglas used in horse trailers should be tinted to reduce glare and heat from the sun. Plexiglas is better in the event of an accident, but scratches more easily than glass. It is best if windows are outfitted with screens to keep out

## *Door Latches and Holdbacks*

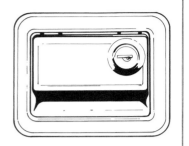

**Slam latch**

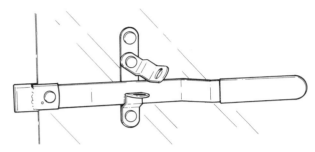

**Rear door latch**

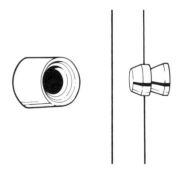

**Rubber holdback**

**Metal holdback**

bugs. Look for windows that can be opened from the outside and latched whether open or closed.

A minimum of one bus-style (sliding) window at the head and one at the tail of each horse is suggested. A third window per horse located on the back door is even better. Sliding bus windows should be located at the rear of each horse in a slant-load, alongside each horse in a straight-load, and in the rear doors whenever possible.

Drop-down windows have smaller sliding bus windows in them, which can be left open for traveling. Although it is fine to open drop-down windows so that a horse can put his head out when the vehicle is stopped, *never* leave them open when traveling unless the opening is also outfitted with a strong interior grille. Some drop-down windows have a positive latch-open feature and an interior grille so the windows can be open for traveling. The interior grilles often have hinges so that they can be lowered when the trailer is at a standstill so the horses can put their heads out.

There should be a minimum of one roof vent per stall with safe, rounded trim on the vents. Because the vents open and close by operating them from the inside, if you are less than 5 foot, 6 inches tall, you might need a step stool to reach the vents in very tall trailers. Two-way vents are best. If you want air to flow in, such as during hot weather, the vent can direct air in through the opening from the front. If you want to draw warm, moist air out of the trailer during cooler weather without creating a breeze in the trailer, then the vent can be opened the other way to draw air out of the trailer toward the rear. The vents should be well fitted and sealed (caulked) to prevent leaking.

## Stall Dividers

Stall dividers can be full and permanent, full and removable, partial and permanent, or partial and removable. Removable stall dividers are handy for hauling large or difficult travelers or for accommodating a mare and foal. Most horses travel more

## Types of Windows

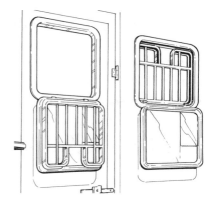

**Drop-down window**

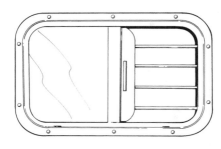

**Sliding bus window**

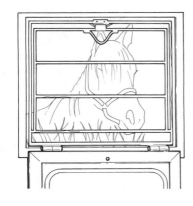

**Homemade grille**

comfortably if the divider goes no more than half-way to the floor. Full dividers that go all the way to the floor tend to make a horse feel trapped and there often is more scrambling to keep balanced.

The head portion of the divider can be either solid or have vertical bars. With a solid divider, there will be less playing or fighting between horses, but bar dividers provide greater airflow for comfort. Slant load dividers usually fold flat against the wall, which is a handy feature for loading.

Vinyl-covered padding on the stall fronts and sides, the center divider, and the back door increases safety and comfort.

### BUTT BARS AND CHEST BARS

A butt bar should be positioned so a horse contacts it before he feels the pressure on his lead rope or hits the back door. Many horses balance themselves by leaning on the butt bar, so the device and its fasteners must be strong and located low enough to support the horse.

Some straight-load trailers have a chest bar instead of a manger. Feed can be placed on the floor, a height that is healthy and comfortable for eating. In addition, the horse can lower his head to blow and clear his respiratory passages en route. The horse's front legs do not bang into a manger wall, and there is no feeder or hay net to get tangled in.

**Removable center divider with permanent center post**

Fastening and releasing the bars should be quick and easy — it is not something you want to have to play with or bang on. Some can be difficult to operate if the trailer is on less than 100 percent level ground. It is best if the bars can be unfastened when there is pressure on them.

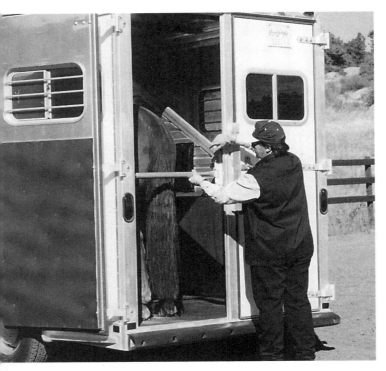

**Fastening the butt bar**

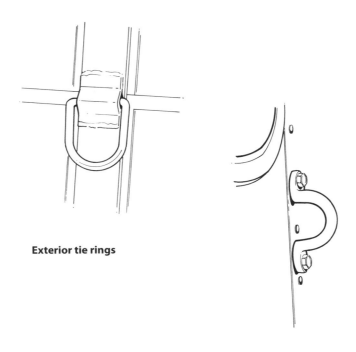

**Exterior tie rings**

### TIE RINGS

Tie rings should be strong and safe, and there should be at least one inside and one outside for each horse. Inside tie rings should be strong and mounted where they can be easily accessed from inside and outside the trailer through a window or door.

Outside tie rings should be located away from door latches, light fixtures, or anything your horse could chew on or become snagged on. Check that they are welded to the trailer framework or secured by bolts that go all the way through the trailer wall.

## Lights

A trailer should have highly visible exterior operating lights and clearance lights, as well as interior lights in at least the manger and tack areas. Interior stall lights and exterior loading lights are useful when loading/unloading or feeding at night. For night traveling, you can reduce the dramatic contrast of oncoming bright headlights for the horses by turning on the interior manger and stall lights. To ensure you'll have lights when you need them, learn which lights have their own separate battery source, which operate when the trailer wiring is plugged into the truck, and which require that the truck's ignition be engaged.

## Tires and Wheels

Tires and wheels must be the proper size for your trailer. A two-horse trailer should have at least Load Range C (6-ply rating); a three- or four-horse trailer should have Load Range D (8-ply rating); four-horse and larger trailers require Load Range E (8–10-ply rating). Make sure tires are inflated to their prescribed pressure and check them often. Radial tires are the preferred choice for a softer trailer ride (see chapter 2 for a discussion of radial vs. bias ply).

Replace trailer tires when they become worn or weather-checked, or if the wear is uneven. A bald tire might make it to the next show or it might put you in an emergency situation along a busy road.

**Exterior lights and recessed light switches for interior lights**

### SPARE TIRE

A spare tire may or may not be included when you buy a new trailer. If not, purchase one and make sure it is mounted on the proper size and style wheel for your trailer. Stow your tire where it can be locked to prevent theft.

# Feeders

Many straight-load trailers have a manger for feeding a horse hay or grain. The disadvantage of manger feeders is that the horse cannot lower his head and neck to blow, cough, and release dust and debris from the hay. The ideal trailer has just a chest bar and room ahead of it that allows the horse to stretch his neck and lower his head in transit. If you use a manger trailer, be sure to only feed very clean hay and unload your horse every four hours to lower his head and blow. In straight-load trailers with chest bars, as well as slant-load and stock trailers, you can use hay nets, hay bags, or other types of feeders.

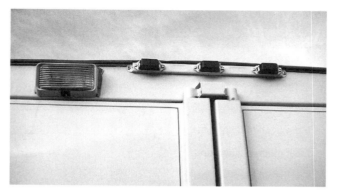

**Clearance lights**

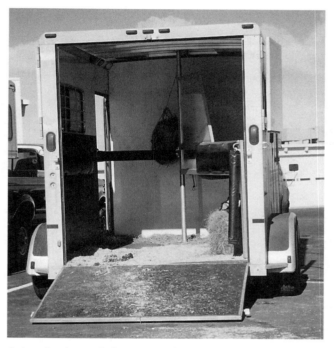

**A chest bar instead of a manger allows a horse to lower his head to stretch and blow.**

# Workmanship

The workmanship and materials dictate, to a large degree, the cost of a trailer. Quality workmanship will be evident in the smoothness and evenness of welds, the fitting of seams, the finishing of edges, and the paint job.

A top-notch paint job not only improves a trailer's appearance but protects it as well. Progressive trailer manufacturers have taken advantage of recent technological advancements in paint quality and bonding. White is the most popular color for

## Trailer Wish List Worksheet

Towing capacity of your truck _____

Straight-pull, gooseneck, or fifth-wheel _____

Steel or aluminum frame _____

Steel or aluminum skin _____

Number of horses (stalls)_____

Size of horses being hauled _____

_____

_____

Required GVW of trailer _____

Stall configuration (straight-load, slant-load, or stock trailer)

_____

Required inside height_____

_____

Required inside length of stall _____

_____

Required width of stall _____

_____

Loading method (step-up or ramp) _____

Tack compartment or dressing room _____

Living quarters _____

Budget constraints _____

_____

New or used _____

both the inside and outside of a trailer and for good reason. White absorbs the least amount of light and heat, and a light interior is more inviting to a horse. Dark colors not only absorb heat, but also make the interior look like a dangerous cave to a horse. You could order a custom red or black trailer to match your truck, but a white trailer would likely be more comfortable for your horse. A white roof is especially important.

All interior and exterior surfaces should be flush whenever possible, and all edges should be finished round or covered with rubber or vinyl edging. Bolts and fasteners should have rounded, smooth heads or be flush with the surface they are on. Smooth interiors from top to bottom will minimize injuries in the event of a trailer rolling in an accident.

# Optional Features

There are many options to choose from that will help you customize your trailer to suit your needs. You might wish to consider a drip cap over the rear doors to send rain to the sides rather than down on you and your horse when you load or unload.

Options that add to a trailer's life and appearance include gravel guards and undercoating. Chip guard, gravel guard, and diamond plate are all types of extra protection that can be put on the front and fenders of the trailer to keep rocks from chipping the paint.

A roof rack makes a handy place to haul covered hay and other large items. Make sure all items are securely fastened to prevent them from blowing or bouncing off onto the roadway or someone else's vehicle. Other features include a spare tire mount, wheel covers, hayrack, ladder, and water tanks.

Now that you have the nuts and bolts when it comes to trailers, you should be able to make a good choice. Use the accompanying worksheet and flow chart to walk through the decision-making process.

# Trailer Selection Flow Chart

**1** **Do you already have a tow vehicle?**
- If yes, go to 2.
- If no, go to 3.

**2** **If you already have a tow vehicle, determine its towing capacity and then use the following guide. If you have:**
- A ¾-ton or larger flatbed or pickup with open bed, consider a gooseneck.
- A ½-ton (full-size) pickup with open bed, consider a two-horse gooseneck.
- A ½-ton (full-size) pickup with camper or large sport utility vehicle or van, consider a two-horse straight-pull.
- An SUV, van, or small pickup that meets minimum towing requirements, consider a one- or two-horse Euro-style trailer.
- Go to 4.

**3** **If you are buying both a truck and a trailer at the same time, unless you are positive you will never want to haul more than two medium-sized horses in a simple two-horse trailer, choose a vehicle with ample towing capacity that will allow you to pull a four-horse gooseneck trailer.**
- Go to 4.

**4** **What will you use your trailer for? If you plan to use it for:**
- Basic economical transport, consider a two-horse straight-pull trailer or a gooseneck stock trailer.
- Camping, consider a gooseneck living quarters horse trailer with a slide-out section and rollout awning.
- Horse shows, economy program, consider a one- or two-horse straight-pull trailer.
- Horse show, luxury program, consider a two- to six-horse gooseneck trailer with or without living quarters.

**5** **How many horses do you plan to haul? If you plan to haul:**
- One horse, consider a lightweight Euro-style trailer or a two-horse trailer with removable divider.
- Two horses, consider a two-horse straight-pull or two-horse gooseneck with or without dressing room.
- Three horses, consider a three- or four-horse gooseneck or a three- or four-horse straight-pull, with or without dressing room.

**6** **What size/breed of horses will you be hauling? If you plan to haul:**
- Horses up to 16 hands, consider a straight-load or slant-load trailer with inside height of 84 inches.
- Horses up to 16 hands-17 hands, consider a straight-load trailer with inside height of 88 inches.
- Horses up to 17.2 hands, consider a straight-load trailer or van with inside height of 92 inches.
- Horses larger than 17.2 hands, consider a custom-built trailer or van.

# 10 Buying a Horse Trailer

Now that you are familiar with many of the choices and features available in horse trailers, you can put together a checklist of the important items and desired options for your particular situation. Whether you are buying a new or used trailer, you need to first know what you want and then gather information about prospective trailers.

Much of the information in chapter 3 about buying a tractor is applicable here. That includes buying from a dealer, at an auction, or by private treaty; deciding to buy new or used; and whether you should take someone along with you on your shopping trips. What's specific to purchasing a horse trailer is whether you are buying a trailer to work with an existing truck or whether you are choosing a trailer and truck together.

Don't assume that just because a trailer looks good, it is safe. If you are new to trailer buying, seek advice from an experienced horseman. Unless you're very knowledgeable, instead of going bargain hunting, you're better off buying a brand-name, well-respected trailer from a reputable dealer that also offers service. Top-of-the-line trailers often carry a five-year warranty. Study the warranty and bone up on trailer and towing vehicle maintenance.

Whether you are buying new or used, you'll want to compare various features of trailers to see how they stack up to your wish list. Just as with tractor shopping, you can narrow the field most efficiently if you follow this four-step process:

- Essentials
- History
- Physical Inspection
- Test-drive

## Essentials

Does the trailer you are looking at have the essentials you require? You should be able to determine this from a classified ad and/or phone call. Make a copy of the checklist for each trailer that you are considering. If a trailer sounds promising, continue with its history in the next section.

## History

You should ask the seller a few things about the trailer to help you determine if it was lightly used or "rode hard and put up wet"! Assuming that your initial questions determine that it has the essentials, these are good follow-up questions to ask when you first call to inquire about the trailer.

- How old is the trailer?
- How long have you had it?
- What have you used it for?
- How many miles do you think you have pulled it?
- Is it in running shape now?
- Why are you selling it?
- Has it been in an accident?
- Has it had any major repairs or renovations?
- What are the trailer's strong points?
- What are its weak points?

### Things to Take

- ❏ Gloves
- ❏ Clipboard with checklists and pen
- ❏ My truck specs
- ❏ My horse specs
- ❏ My gear specs
- ❏ Camera
- ❏ Brochures
- ❏ Dime
- ❏ Tape measure
- ❏ Flashlight

### Things to Know

**MY TRUCK SPECS**

Wheelbase _____

MTWR (look on plate inside driver's door or truck brochure)_____

GVWR_____

Tailgate height clearance needed for gooseneck trailer _____

Estimated GVW when loaded for traveling _____

_____

Maximum weight of a loaded trailer that can be safely pulled by my truck _____

**MY HORSE SPECS**

Height at withers _____

Height of head when alert _____

Length from chest to tail_____

Length from nose to tail _____

Width at widest part of barrel _____

(Figure several inches clearance in all directions when measuring horse trailer spaces.)

**MY GEAR SPECS**

Size of footlocker or tack trunk _____

Size of water caddy _____

Number of saddles, bridles, other tack _____

Other gear such as clothing, cooler_____

## Trailer Essentials Fact Sheet

Date _____

Seller's name and contact information _____

_____

_____

_____

_____

Make and year of trailer _____

_____

_____

Hitch type and rating _____

_____

Hauling capacity (number of horses) _____

_____

Ramp or Step Up_____

GVWR_____

_____

Configuration _____

_____

Height_____

_____

Width _____

_____

Brakes on all wheels _____

_____

Breakaway safety brake_____

_____

Windows _____

_____

Proper venting _____

_____

Tie rings _____

Tack storage _____

_____

Dressing room, if applicable _____

_____

Living quarters, if applicable _____

_____

Floors and mats _____

_____

Lights, interior_____

Lights, exterior _____

Per seller, items that need repair to bring this trailer up to
safety and operational status (replace floor boards, brake
job, or other faults) _____

_____

_____

_____

_____

_____

Other notes_____

_____

_____

_____

_____

_____

_____

_____

# Physical Inspection

If a trailer sounds promising after the first two stages, make an appointment to look at it. Here is where you want to spend some time looking at details. If you are buying a used trailer, take your tow vehicle along so that you can hook it up to evaluate the hitch or bed height, assess levelness, and check the lights and brakes.

Used trailers generally have no warranty, so you or your mechanic should inspect the trailer carefully.

## OVERALL APPEARANCE AND CONDITION

Examine the interior sidewalls, fenders, and front end for damage, rust, and sharp edges. Fresh paint on a trailer might be hiding rust, which compromises the strength and safety of the trailer. Look for paint bubbles or a bumpy texture that might indicate rust under the new paint. Conversely, if the trailer needs a coat of paint, note whether it would need sand-

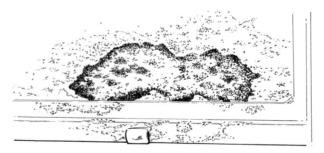

**Rust along the lower part of this door indicates that the door has had manure and urine packed against it for long periods of time.**

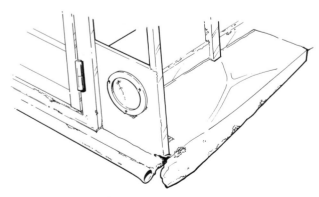

**The dangerous sharp, broken edge along the bottom of this trailer is hazardously to a horse's legs.**

blasting to remove rust or bodywork to repair dents before painting. Dark-colored trailers are uninviting during loading and hot during travel, so realize that if you buy a dark trailer, you will probably end up repainting it.

## STRUCTURAL INSPECTION

Check the axles and frame for excessive rust, bending or warping, or signs of welding or other repair. Any damage to the frame requires costly repairs, so if you find major problems, look for another trailer.

Check for broken springs and worn bushings where the springs attach to the frame.

Inspect the roof from the inside and from the top for dents, cracks, or other damage, and for signs of leaking.

Note how the wall lining is attached to the frame work and if there is any looseness or rust, both of which can indicate that moisture has seeped into the walls. They could be structurally unsound.

### Serial Number and Information Plate

The trailer's vehicle information plate is usually located on the side, the tongue, or inside the dressing room of the trailer. It will contain the following information:

- ❏ Manufacturer's name and location and the date of manufacture
- ❏ Serial number
- ❏ Model name and number
- ❏ Year of manufacture
- ❏ GVWR (gross vehicle weight rating)
- ❏ GAWR (gross axle weight rating) for both the front and rear axles
- ❏ Empty trailer weight
- ❏ Tire size
- ❏ Maximum tire pressure (per square inch), cold
- ❏ Maximum weight-carrying capacity per tire
- ❏ Rim size and number of holes
- ❏ Maximum weight capacity per rim

## Physical Inspection Checklist

### STRUCTURE

Stored indoors or outdoors? _____

Rust _____

Paint _____

### BODY

Frame _____

Suspension _____

Roof _____

Ramp (raise and lower) _____

### DOORS

Hinges _____

Latches _____

### WINDOWS

Hinges or slides _____

Latches _____

Vents (number and location) _____

### STALLS

Verify height, width, and length _____

Floor (material; condition) _____

Mats _____

Sidewalls (reinforcement?) _____

Dividers _____

Padding _____

Butt bars and chest bars _____

### HITCH

Coupler _____

Safety chains _____

Jack _____

Wiring harness/plug _____

Breakaway switch and cable _____

Brake battery _____

### TIRES AND WHEELS

Size and condition of tires _____

Spare tire _____

Axles _____

Wheel bearings _____

Brake shoes and drums _____

### TACK ROOM

Condition (clean; musty smelling?) _____

Hooks/racks _____

Carpet _____

### MISCELLANEOUS

Operating lights and covers

    Brake _____

    Back-up _____

    Turn signal _____

    Running/clearance _____

    Interior _____

    Loading _____

License plate holder and light _____

Tie rings (number interior and exterior) _____

### UNDER THE TRAILER

Rust _____

Damage _____

Wiring _____

Floor (from underside) _____

### PAPERWORK

Title _____

Maintenance schedule and log _____

Repair records _____

Owner's manual _____

Safety warning stickers _____

If the trailer has a ramp, raise and lower it a few times to see how well it functions and if it is a one- or two-person ramp. Check the floor of the ramp to be sure it is sound.

All doors should swing freely and hook open using whatever type of holdback mechanism the trailer has. Check all hinges and latches of all doors to be sure they are working and don't need repair or replacement. Again, be very vigilant for rust damage. If the doors have protruding latches, determine if they can be replaced with safer ones or if you can live with the risk of injury involved.

Be sure all windows operate properly. Open and close each window and vent and note any signs of damage or leaking around them. Be sure all hinges are fully operational and that all handles and latches work. Assess the latches for safety as with the doors. Make sure the ceiling vents operate properly and are not rusted shut.

Old-fashioned butterfly side vents on manger doors can be operated from the outside of the trailer, which is handy, but in hot weather, they don't do much to remove warm, moist air from the trailer. In cold weather, they deliver a chilling wind onto the horse. Strategically placed ceiling vents are the best.

## CONDITION OF STALLS

Bring your tape measure and take accurate measurements of inside and outside height, width, and length. Remember, most older model horse trailers are narrower and shorter (inside height of 78 inches or less) because they were made for horses about 14.2 hands tall. For a 16-hand horse, look for a stall that is 32–34 inches wide and 80–84 inches tall.

Mats should have some cushion left and not be all chopped up from hooves. Investing in new mats is reasonable if the trailer is worth buying otherwise.

Remove the mats and check the floorboards for signs of moisture, rotting, splitting, or warping. How aggressive you can be with stomping and poking (with a screwdriver) will depend on what the seller allows, but even if you jump up and down as hard as you can, it won't come close to the force of a 1200-pound horse (or two) on a bumpy road, so the floor must be sturdy and solid. Often if a trailer floor is shot, the lower sidewall attachment to the frame work is probably on the way out, too.

Good ventilation is essential for your horse's health. These small butterfly vents are inadequate and poorly placed. Ceiling vents are better.

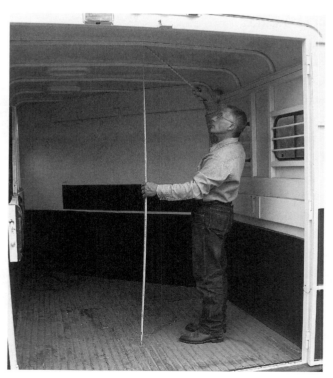

Using your spec sheets, compare your horse's numbers with the interior dimensions of the trailer to see if it will work.

**No matter how cheap, some trailers aren't a bargain. This stock trailer is worn out and originally designed for cattle.**

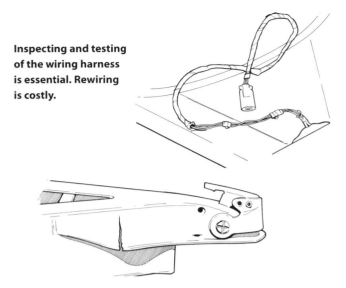

**Inspecting and testing of the wiring harness is essential. Rewiring is costly.**

**The hitch must be sound and reliable. This cracked coupler could break off at the next bump.**

Check the sidewalls for scarring or dents from scrambling hooves. If damaged, the lining on steel trailers could be buckling away from its attachment at the floor, creating a dangerously sharp edge for the horse's legs. Most steel trailers eventually need side-wall reinforcement or replacement.

Check that the trailer has rubber-reinforced side, rear, or front walls. Inspect the padding on the sides, front and rear. Look at the condition of the mangers or eating area.

## HITCH INSPECTION

Check the hitch, coupler, and chains for cracks, wear, warping, or welds indicating repair. Make sure everything is there; you'll check it again during your test-drive. Account for all chains, plugs, cables, and bars. Note if wires are bare or have cracked insulation.

Trucks have been getting taller, especially because of the popularity of four-wheel-drive vehicles. Many trailer manufacturers have made adjustments to their trailers to accommodate the higher beds so that their trailers will clear the bed rails and tailgates of taller trucks. The only way to determine if a truck and trailer will match is to hook them up and see.

## WHEEL BEARINGS AND TIRES

Is there grease leaking out of the bearing caps or behind the bearings? If so, it could mean the seal is shot. Remove one or all wheels and brake drums and inspect the bearings, brake pads, and drums.

Don't just kick the tires — inspect them closely for weather checking (signs of cracking or rotting), which could indicate the need to replace them. The sidewalls of a stored trailer tire deteriorate from sun and moisture before the treads do, so check carefully. Also look for uneven wear on individual tires, which may indicate frame or axle damage or faulty wheel bearings.

Look for tires with a minimum depth of a ¼-inch tread (about one-third the diameter of a dime). Be sure the tires are the proper size and rating for the trailer. Ask about their age and mileage. Tires can be replaced, but factor in the cost.

Is there a spare tire? Where is it stored? What condition is it in? Can it be locked up?

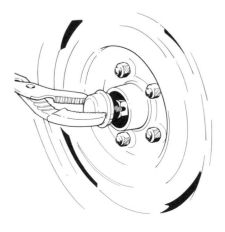

**Remove the dust cap to check wheel bearing grease and to access the nut that holds the wheel on.**

The severe weather checking on this trailer tire means it is riding on a hope and a prayer.

Good tread on the left; no tread on the right.

Look for tires with tread depth at least one third the diameter of a dime.

### UNDER THE TRAILER

Crawl under the trailer to examine the frame and axles and check for cracks, corrosion, loose bolts, loose wires, or anything that looks broken or missing. Compare one side to the other. Check the floor from the bottom to assess its integrity. Follow brake wiring from the plug to the brakes and lights, making note of any damage, repair splices, or cracked conduit.

### CONDITION OF TACK ROOM

Does the tack room have a fresh smell, or is it musty and moldy? Inspect the ceiling, walls, and floors for signs of leaks. If the tack room is carpeted, note its condition. Be sure the door and windows are fully operational. Count the number of hooks, saddle racks, and other tack storage devices. Note if the lights work.

### MISCELLANEOUS

Note what kind of lights are on the trailer. You'll test them later. You must have brake lights, taillights, and a minimum of running lights, but additional clearance lights and interior stall and tack room lights are desirable.

Where is the license plate holder located? Is it in a safe place where a horse cannot contact its sharp edge? Is there a frame, or is the plate mounted bare? Is there a light, and does it work?

Tie rings both inside and outside the trailer should be of the proper height, solidly attached, and show no signs of damage. Although tie rings can be replaced or repaired, it requires a savvy craftsman to make them as safe and strong as needed.

Safety warning stickers near wheels and the hitch should be present and legible.

### PAPERWORK

Make sure the seller has a clear title to the trailer.

If you are looking at a reconditioned trailer at a dealership, find out what the warranty covers. If the warranty of a newer trailer is transferable from the original owner, find out what it covers and how much time is left. Most used trailers have no warranty, so you or your mechanic should look a prospect over very thoroughly before buying it.

Ask to see the trailer maintenance log or receipts. Although they don't prove anything one way or another, it will help you determine if the seller has followed a maintenance plan.

Ask if the seller still has the owner's manual, which will provide you with a lot of specific information related to the construction specifications and maintenance of the trailer.

# Test-drive

If your towing vehicle is set up to pull the trailer, perhaps the seller will allow you to tow it down the road for a test-drive. Otherwise, you might ask the seller to hook the trailer to his truck and demonstrate its roadworthiness. Here are some factors to note:

- Ease of hitching
- Levelness
- Condition of plug and wire
- Performance of brakes
- Condition of breakaway cable
- Straightness of towing
- Items needing repair or replacement

## EASE OF HITCHING

The trailer jack (jacks on a gooseneck) should raise and lower the trailer smoothly with minimal effort on your part. When the trailer is hooked, the jack(s) should remove easily or raise up high enough to be out of the way when traveling.

## ON THE LEVEL

The floor of the trailer should be level when the trailer is fully hitched to the truck. If the trailer is elevated or dropped in front, you'll need to use a stepped ball mount to get it level. With a gooseneck or fifth-wheel, see if there is enough adjustment in the coupler or hitch to make the trailer level.

## WIRING HARNESS

Make sure the insulation on all wires is intact with no bare wires showing. The trailer plug should match the truck socket without trying to force it in place. The contact holes in the plug should be relatively clean and not filled with dirt or corrosion.

With the trailer coupled and plugged to a tow vehicle, test the running lights, turn signals, brake lights, and interior lights. If a light doesn't work, and the owner claims it is just a bad bulb, ask him to replace the bulb so you can test it. If it turns out it isn't just a bulb, it could mean costly wiring troubleshooting and repair.

## BRAKES AND BREAKAWAY CABLE

While slowly towing the trailer forward, operate the truck's auxiliary brake controller and have someone watch to see if all the trailer wheels lock or at least provide significant resistance. The breakaway cable should be smooth and intact with no broken strands sticking out. It should have a loop or a snap on the end for attaching easily to the tow vehicle.

Open the battery case and check that the connections are not corroded. If you have a battery tester, check the battery for voltage. If the owner agrees, pull the breakaway cable to activate the mechanism and then tow the trailer forward a few feet to see if the battery is strong enough to lock the brakes.

## TOW STRAIGHT

Ask your assistant to follow behind you in another vehicle and look for any wobbling of the wheels (loose lugs or shot bearings), any swaying of the trailer (low tires or imbalance), any pulling to the left or the right on braking (brake adjustment), or an overall crablike movement of the trailer where it seems to travel at a diagonal down the road (bent frame). Some of these things, like loose lugs or low tires, can be remedied fairly easily; others, like a bent frame, are real deal breakers.

Narrow the field down to two or three prospects and then have your mechanic give you the go-ahead before you plunk down the cash.

# Safety and Maintenance

**N**ow that you have that ¾-ton truck and trailer, as well as your dream tractor and attachments, you might feel invincible with all that power and steel surrounding you. That feeling lands a lot of people in trouble. These are not toys, but potentially dangerous pieces of equipment. Use them responsibly for your safety, your horse's safety, and the safety of others on the road or farm. Learn how to properly maintain your equipment as a matter of both safety and economics.

Operating farm equipment can be dangerous. Tractors, APVs, and trucks are not for children to play with or for adults to operate carelessly. Inform every person who operates equipment on your farm about basic safety practices and make sure they agree to follow safe operating procedures.

# Safety

Safety features are there for a reason. Keep all protective equipment in place and be sure all guards, shields, and safety signs are installed properly and maintained. If you remove, modify, or disengage safety features, you increase your risk of injury or death. Guards are there to protect you from injury from such things as the power take-off, chains, belts, and universal joints.

Whenever you use your tractor or trailer, check for children, animals, and other bystanders before you start moving.

Sit properly in the driver's seat with your seatbelt fastened.

Dress appropriately for the task at hand — use protective footgear, gloves, and goggles when necessary. Loose clothing, jewelry, and long hair can be caught on moving parts.

Know your equipment and its capacity. Learn to hitch and unhitch implements properly and safely, and take your time to do the job correctly.

Keep the cab and the floor of your tractor and truck clear of dirt, snow, ice, and debris so you have the best footing possible when driving. Keep the foot pedals clean and in good repair. Don't use the floor to store tools, lunch pail, thermos, chains, ropes, and so on. People get hurt tripping or slipping on such items all the time, and loose objects can slide under a brake pedal just when you need to stop. Instead of stashing stuff hither and yon, install suitable storage boxes and racks to hold all the items you need to take with you.

## TRACTOR SAFETY TIPS

Statistics show that more accidents and deaths are attributed to machinery that is used in agriculture than in any other industry in the United States. Of those agricultural accidents, tractor incidents account for 70 percent, with an estimated 350–450 fatalities per year. The most common tractor accidents are:

- Driver falls off when mounting or dismounting or after hitting a bump while speeding
- Operator becomes entangled in the PTO shaft while it is running
- Tractor rolls over when unbalanced, on uneven ground, or when turning sharply
- Tractor flips over backward due to improperly hitching a load that is too heavy or one that is attached above the tractor's center of gravity
- Tractor runs over operator or bystander

It is your responsibility to know and obey all local, state, and federal laws related to your tractor and equipment. Here are some tips on safe tractor operation.

*Be aware of the importance of weight and balance.* Have an expert evaluate if your tractor is safe to operate with your implements. Add frame weights, tire weights, or tire fluid if necessary to balance your tractor.

*Hitch only to the drawbar when pulling a load.* Hitching to the axle housing, seat base, top link of the three-point hitch, or any point above the drawbar reduces pulling capacity and increases the chance of the tractor flipping over backward. Be sure no loose chains, ropes, or cables are dangling or dragging from either the tractor or the implement. They can catch under a wheel, on a stump, or on a rock and cause a serious or fatal accident.

*Mount with three-point contact.* Make contact with three of your limbs (arms and legs) as you mount and dismount. Do not grab the controls to pull yourself up. Always start the tractor from the driver's seat, never while standing off to one side or from the ground.

*Do not carry passengers on the tractor or implements.* Most tractors only have one seat and are not built for passengers so there is rarely a safe or comfortable place for a passenger to ride. Passengers often obstruct an operator's view and access to the controls.

*A tractor is a workhorse, not a racehorse.* Drive at safe speeds; reduce speed before turning to prevent overturning. Reduce speed when crossing slopes or when driving on rough, muddy, or slick surfaces.

Using a three-point mount, you will always have contact between three of your limbs and the tractor.

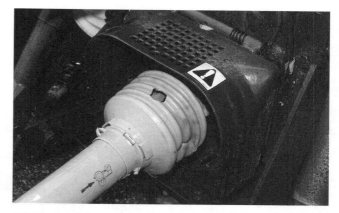

**Shields are a mandatory safety feature. Older tractors should be retrofitted.**

**To minimize the risk of tipping, keep the bucket low, especially on slopes and rough terrain.**

*Know the terrain where you will be driving.* Walk unfamiliar terrain to identify ditches, large rocks, debris, or bogs. When driving in hilly terrain, keep the tractor in gear — never coast downhill or depress the clutch while going uphill or downhill. Drive straight down all slopes, never diagonally. If you must turn on a slope, turn downhill. When you have to drive up a slope, consider backing up for added safety.

*Never attempt to service your tractor or implements while they are running.* Stay clear of all rotating PTO implements and make sure other people and animals are clear of them, too. Remember, the PTO makes nine revolutions every second, so it takes just a fraction of a second to become entwined. Never step over or lean over a running PTO. Before disconnecting hydraulic lines, release the pressure.

When you stop the tractor and plan to leave the driver's seat:

1. Come to a full stop.
2. Disengage the PTO.
3. Lower implements and attachments to the ground.
4. Put the transmission in neutral or park.
5. Apply the parking brake or lock the brakes.
6. Shut off the engine.
7. Remove the key if tractor will be unattended.
8. Block the tires if necessary.

## ROPS Safety

Use a seat belt in a tractor equipped with a rollover protective structure. If your tractor overturns, hold the steering wheel firmly and don't leave the seat

A ROPS system consists of a rollover protective structure and a seat belt.

When driving a tractor on the road, use a slow-moving vehicle sign and good judgment.

until the tractor has come to rest. A seat belt and ROPS are estimated to be 99 percent effective during a tractor rollover — if your seatbelt is properly fastened, your body will be held in place and protected by the ROPS.

However, it is generally recommended that drivers not use a seat belt when operating a tractor without a ROPS. Not being restrained by a seat belt makes it possible for you to jump or be thrown clear if the tractor overturns, instead of being trapped or crushed beneath it.

### On the Road with a Tractor

Before driving your tractor on a public roadway, make sure all lights work, brakes and steering are in good working order, and the ROPS is in place.

Attach a slow-moving vehicle emblem to the back of your tractor or implement. If legal, use flashing amber warning (hazard) lights; in some areas, use of hazard lights when driving is illegal.

Towed implements should be raised and locked in transport position with a safety hitch pin and chains in place.

### TRUCK AND TRAILER SAFETY TIPS

When you travel with your horse, the number one priority is for both of you to arrive safely. For peace of mind, develop the habit of preparing thoroughly and safely for every trip, using detailed checklists. Develop and practice good trailer driving skills. Follow up with post-travel care for your truck and trailer so they are ready for the next trip.

*It is a good idea to make a spare set of keys for ignition, glove box, tonneau or camper, spare tire lock, trailer hitch lock, tack room doors, trailer windows, and anything else that has a lock on it.* Keep a complete set of spare keys in a safe, accessible place in the towing vehicle. In addition, put a master vehicle key in a magnetic box in a protected spot on the exterior of your tow vehicle.

*Keep the bed of your truck clean to keep debris from blowing around and flying at your trailer.* This is essential if you are towing a gooseneck so that the hitch, chains, and cables will operate freely and not hang up. If a loose object yanks the breakaway cable free, you and your horses could have a serious accident.

*When your trailer is parked, use one or two heavy wheel blocks on each side of the trailer in between the front and rear wheels to prevent it from rolling.* If you only block one side, the trailer can pivot and roll off the jack wheel. If you park on dirt, put a flat board (at least 12 inches square) under the wheel of the trailer jack to allow you to move the trailer somewhat when positioning it to hitch up. The board will also prevent the jack from sinking into soft ground or getting frozen in the mud. A wheel chock, with a concave surface for the wheel to fit in, works just as well for supporting the jack wheel, plus it helps keep the trailer from moving.

### APV SAFETY TIPS

All-purpose vehicles can be useful to introduce youngsters to the skills and responsibilities of operating motorized vehicles. But even though these vehicles can be exciting and fun to drive, they should always be viewed as tools, not toys. Here are some additional guidelines from the ATV Safety Institute that apply to both all-terrain vehicles and utility task vehicles:

- Take an approved training course.
- Ride an ATV that's right for your age:
    Age 6 and older: less than 70cc engine
    Age 12 and older: 70cc–90cc engine
    Age 16 and older: more than 90cc engine
- Provide competent adult supervision of riders younger than 16 years of age.
- Do not ride unlicensed vehicles on public roadways (many states will not license ATVs or UTVs; check with your local Department of Motor Vehicles).
- Do not drive when upset or when using alcohol or drugs.

- Do not wear headphones when driving.
- Keep speed suitable for the weather conditions and terrain. Just because you *can* go fast doesn't mean it is safe to do so.
- Be aware of protecting the environment.
- Never carry more persons than the vehicle is designed for.

# Maintenance

You have made a big investment in the equipment needed to run your property. If you take care of your vehicles, they will last longer and be more reliable.

## Snowblower Safety

Because of the nature of the work and the conditions in which it is used, a snowblower can be very dangerous.

- Always use tire chains on the vehicle that is operating a snowblower.
- Before the first snow, remove all loose objects, such as stones, toys, and dog bones, from areas where the snowblower will be used.
- Place stakes or flags to mark the location of objects such as banks and ditches, large rocks, small hydrants, stumps, well casing, and small bushes that might be hidden by snow.
- Set the skid plates to ride at least one inch above the ground when clearing graveled areas.
- Direct snow discharge away from animals, buildings, and vehicles.
- Don't wear loose, long clothing — especially dangling gloves or hanging scarves — that might get caught in the auger.
- Drive straight up and down a slope, never across, because your vehicle could easily slip or tip over.

Tip: Spray the chute with silicone lubricant before starting to minimize clogging and increase throwing distance.

Storing your implements indoors will greatly increase their life and resale value.

Each time you use your trailer, perform a before-use check and an after-use follow-up. Fix or service anything that warrants attention at those times. Perform annual maintenance procedures as outlined in your owner's manual or as suggested in this chapter. Store and secure your trailer properly. Regularly go over your equipment with a fine-toothed comb, much as you would a used tractor or trailer that you were considering buying.

## TRACTOR MAINTENANCE

It used to be that most people with a modicum of mechanical ability could fix a lot of things that broke on a tractor. But like autos, tractors are becoming more complex all the time, and what used to be a simple repair now often requires considerable skill and training to tackle and can be very expensive. That's why it is always best to spot little problems before they become big ones. Just as you give your horse or your tack a daily check, walk around your tractor on a regular basis to identify any loose, worn, broken, or missing parts.

In addition to the routine maintenance schedule outlined in your owner's manual, you should regularly check the fluid levels, which are the lifeblood of your tractor. Fluids include radiator coolant, engine oil, transmission fluid, front and rear differential and transfer case oils, and hydraulic oil. Many of these have dipsticks that make it easy to see fluid level at a glance. To check other fluids, however, you'll need to use a wrench to remove a plug so that you can check the level visually or by sticking your pinky in the hole to feel the level.

Most fluid levels should be checked weekly. During hot weather, however, check radiator coolant level daily. It should be one inch over the top of the radiator core. Cooling systems that use an overflow tank can be checked by looking at the tank without removing the radiator cap. Never remove the cap from a very hot radiator, as the escaping pressurized steam can cause serious burns.

If you are operating in very dusty conditions, you should also check the air cleaner daily. Like your horse, your tractor needs plenty of clean air in order to perform well.

Each time before operating your tractor, look closely for signs of oil leaks on the engine, transmission, and axles. Correct leaks as soon as possible. A simple leak detector is a piece of clean cardboard placed under your tractor when you park it. The smallest drop of fluid will leave its mark and help you find an early leak before it becomes serious.

## Battery Cautions

Sulfuric acid is hazardous to your health and can produce burns and other permanent damage if you come into contact with it. It will also burn holes in clothing. Always wear eye protection and rubber gloves when working with batteries, and it's a good idea to wear old clothes as well.

Keep metal tools and jewelry away from the battery; if you should make contact between the two terminals or between the positive terminal and the vehicle you could be seriously injured, vehicle components could be destroyed, and the battery could be damaged or explode.

**Warning:** Do not expose a lead-acid battery to a lit cigarette, sparks, or flames — flammable gases from the battery could explode, drenching you with sulfuric acid.

## Diesel Fuel

Diesel fuel systems are more finicky than gasoline systems. Diesel vehicles are notoriously hard to start in cold weather, as the fuel will gel or wax at temperatures below 30°F, causing restrictions in fuel lines and fuel filters. At lower temperatures, diesel fuel particles can freeze, especially if the system contains any moisture.

Add a diesel fuel anti-gel product to prevent cold-weather problems. Only a very small amount, about one-half ounce per five gallons of fuel, is needed to keep the fuel flowing freely. Most new models have a glow plug option (or standard installation) that preheats the cylinders for easier starting.

**Warning:** Do not use spray ether in combination with glow plugs to start a diesel; the engine could explode. Never mix gasoline, ethanol blends, or alcohol with diesel fuel because it can damage the engine.

Your operator's manual should contain a diagram identifying the location of all grease zerks.

Remove debris from the surface of the radiator with compressed air and a soft brush.

Sometimes hydraulic and diesel fuel leaks occur under high pressure and these might require special tools to diagnose and repair. High-pressure leaks can be tiny and hard to see. It is not a good idea to feel with your bare hand for leaks in high-pressure lines, because fluid under pressure can penetrate skin or eyes, causing serious injury. Instead, pass a clean piece of cardboard or wood over suspect areas and look for evidence of spray. Always wear safety goggles.

While operating your tractor, pay attention with all of your senses — listen for hisses, rattles, squeals, or whines that are out of place; look for steam, or for loose or rubbing parts; smell for smoke or hot fluids; feel for unusual vibrations or changes in how controls respond; taste that cup of coffee or soda, but don't taste alcohol when operating your tractor!

### Routine Tractor Maintenance

Perform all routine maintenance as prescribed by your owner's manual or by the tractor manufacturer. Here is a general idea of what you need to do.

*Change the engine oil and filter approximately*

*every 150 hours or as specified in your operator's manual.* Be careful if draining oil from an engine that is hot from running, because hot oil could burn your hand when you remove the drain plug. Disposable rubber gloves are a good idea for this procedure. Catch the waste oil in a bucket or tub and dispose of it properly at a recycling center or other facility.

*Change transmission oil and power steering oil and filters approximately every 300 hours or as specified for your tractor.* Use the same interval for front axle oil in four-wheel-drive tractors. Each system on your tractor uses a specific oil, and the oils may not be interchangeable. Make sure you are using a lubricant specified or approved by the manufacturer.

*Grease all fittings every 50 hours or as specified in your operator's manual.* Some grease zerks are easy to spot, while others can take a bit of searching and may entail removing protective covers or rubber plugs. The manual will have a diagram of where to find all the fittings.

# Battery Maintenance and Cleaning

Here are ways to prolong battery life:

♦ Keep it fully charged. A starter battery is usually charged by the vehicle's alternator immediately after use, while electric car batteries should be charged within 24 hours of use. A fully charged battery will withstand freezing to minus 77°F, but a fully discharged battery will freeze at approximately 20°F. *Do not* jump or boost a frozen battery if the case is cracked or until the battery has been fully thawed out, recharged, and tested.

♦ Keep electrolyte level up; battery plates exposed to air immediately begin losing their effectiveness.

♦ Don't deep-cycle a starter battery; quit using it at the first sign of battery weakness and recharge it fully before using it again.

♦ Keep it cool. Batteries lose their charge and deteriorate more quickly when they are hot, so park the vehicle in the shade, not in direct sun.

♦ Charge batteries at least once a week. Batteries naturally lose their charge over time, and a low battery will deposit lead sulfate crystals on its plates. With a starter battery you can simply drive the vehicle for 15 minutes to charge it. Deep-cycle batteries in an electric vehicle need to be plugged into an appropriate charger.

Like other parts of a vehicle, a battery needs regular attention if it is to last a long time. Since a battery has no moving parts, maintenance is fairly simple: Keep it clean and filled with water. Here's how. (Use rubber gloves whenever you handle a battery.)

1. Look over the battery carefully for defective cables, loose connections, corroded cable connectors or battery terminals, cracked cases or covers, loose hold-down clamps, and deformed or loose terminal posts. If you see cracks in the case or obvious damage to the terminals, the battery should be replaced even if it is performing well.

2. If the battery has removable caps on top, remove them (sometimes the caps are grouped into bars). If you have a sealed battery, skip step 3 and go to step 4.

**The corrosion of these golf car batteries prevented them from working.**

3. Look into each cell and add water if necessary to bring the liquid level one-half inch above the tops of the lead plates. *Do not* overfill, because the electrolyte will expand when it gets warm and could overflow. Use distilled, deionized, or demineralized water, or rain water. Tap water and reverse osmosis water from residential systems might contain calcium or magnesium, which can promote sulfation. Be careful when filling that acid doesn't splash out onto your skin or clothes. Replace the caps.

4. If there are powdery deposits on the terminals, they must be cleaned away, because the corrosion blocks the flow of electricity. First remove the cable from the negative terminal (marked with a minus sign) by loosening the nut and using a battery cable puller. If you don't have this tool, you can wiggle the cable until you can lift it off the post, but avoid subjecting the battery terminals to excessive bending or twisting forces.

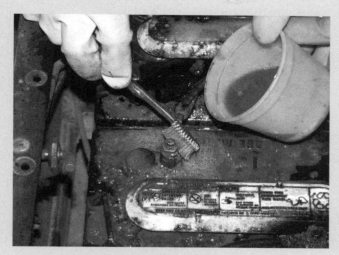

**A baking soda solution and old toothbrush are all that you need to clean off the corrosion.**

Sometimes you can use pliers to spread the end of a clamp to loosen it from the terminal.

5.  Remove the cable from the positive terminal (marked with a plus sign and sometimes colored red). If you try to remove the positive cable first and the wrench touches the vehicle frame, you could damage engine components or blow up the battery. There is also the risk of getting a spark and voltage spike as the positive cable leaves the terminal, which could damage electronic engine components. It's better to disconnect the negative cable first.

6.  Examine the battery cables and clamps to see whether they are frayed or badly corroded and replace them if needed.

7.  Tie the cables out of the way so they don't flop around and touch the terminals.

8.  Scrub the deposits off the terminal posts and cable clamps with an old toothbrush or disposable nonmetallic brush and a mixture of baking soda and water — the ratio is not critical. As long as the baking soda solution bubbles when it contacts a surface, there is still corrosion to be removed. Another method is to rinse the terminals and clamps with hot water while scrubbing. Clean corrosive deposits from the battery sup-

port tray and hold-down parts as well. **Warning:** Do not get baking soda inside the battery because it will neutralize the battery acid and could ruin the battery.

9.  The battery terminals and the inside of the cable clamps should be shiny. If they need further cleaning, use a special round battery terminal cleaner brush. It has one tapered brush that cleans the inside of the clamps and another cylindrical brush that polishes the terminal posts. Alternatively, you can use a soapless steel wool pad or nylon pad.

10. Rinse everything with clean water.

11. Dry everything with a clean, disposable, lint-free rag.

12. Protect the terminals from further corrosion. Corrosion deposits are caused from gases escaping from the battery and condensing on metal parts. To prevent corrosion and deposits, spray the clean terminals and clamps with a battery protector product (available where auto parts are sold) before connecting them. Alternatively, put a small bead of silicon sealer around the base of the post and place a felt battery washer over it. Coat the washer with grease or petroleum jelly.

13. Connect the positive cable first.

14. Connect the negative cable.

15. Spray the terminal connections with a battery protection product or coat them with grease.

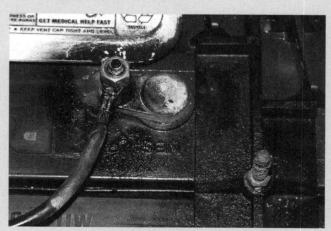

**This happy battery terminal has been cleaned, rinsed, dried, and sealed and is now functional.**

## Antifreeze/Coolant Warning

Ethylene glycol, the most common antifreeze/coolant, extends the freezing and the boiling point of water and also inhibits rust. Even though pure ethylene glycol will freeze at 0°F, when it is added to water it inhibits the formation of ice crystals and thus lowers the freezing temperature of the solution. Most engine manufacturers recommend a 50/50 solution of water and ethylene glycol, which protects to minus 34°F. For colder climates, you can increase the mix to a maximum of 67 percent antifreeze, which protects to minus 84°F.

A stronger solution will actually reverse the benefits. Because a solution containing more than 60 percent antifreeze can lead to clogging of the cooling system, it's best to keep it between 40 percent and 60 percent, if possible. Don't neglect to winterize your tractor, because the damage that occurs when an engine block freezes and breaks is very costly to repair.

**Warning:** Ethylene glycol is toxic. It is doubly dangerous because it has a sweet flavor that makes it attractive to children and animals. Handle it and dispose of it like the hazardous chemical that it is. Do not run it on the ground or flush it down a drain. There are other types of antifreeze available, including propylene glycol, which is less hazardous. Follow the tractor manufacturer's instructions for type and use and never mix different types.

*Drain and flush the radiator annually.* The radiator cap should have a working spring valve and a sound rubber gasket. If you see that the valve is stuck or the gasket is torn, replace the cap with one that has the same pressure specifications as the old one. Use a soft brush and compressed air to remove any debris on the front of the radiator or grill so that airflow will not be obstructed and the radiator can effectively do its job. Shine a light from behind the radiator to check that air passages are free.

*Check fuel hoses for rubbing and leaking and check the clear fuel filter sediment bowl.* It should contain pure clean fuel, with no particles floating or settled on the bottom. If the liquid has layers like vinegar and oil dressing, the bottom layer will be water, which should be removed.

*Check all hoses, belts, and clamps.* Visually inspect and feel all belts and hoses and check all hose clamps for tightness. Replace radiator hoses that are cracked or feel overly soft or hard. A very soft hose has likely been exposed to oil or grease and is dangerous because it can rupture or swell under pressure. A very hard hose has probably gotten too hot from engine heat, the most common cause of hose failure. A brittle hose can easily crack or break.

A common problem with belts is incorrect tension. Too little tension is worse than too much because the belt can slip and overheat. As a rule of thumb, a belt should flex under firm hand pressure about one-fourth inch per foot of belt. Another guideline is to look for belt play of about one-half inch between pulleys. Once you've felt a few properly tensioned belts, you'll be able to tell if a belt is too loose or too tight. Twist each belt with your fingers to inspect the underside as well as the topside. Replace any belts that are shiny, cracked, stained with oil, or show internal cords.

*Remove and clean air filters and replace as needed, usually at least every 50 hours.* The best way to clean the filter is to blow compressed air through the paper filter material, preferably outdoors. You can tell if a filter needs replacing by holding it up to the sun or a bright light — if very little light comes through, then very little air can get through either.

*Check and adjust the brakes regularly and ensure they are evenly adjusted.* You can elect to have your local tractor dealer do this for you, but if you observe the procedure once or twice, you might decide you can do it yourself.

*Look over tires closely for checks, cuts, bulges, and correct pressure.* Service or replace tires as needed. Keep tires at optimum inflation, which is less than the maximum inflation listed on the tire's sidewall. Check tire air pressure every couple of weeks.

*Check for signs of corrosion on battery terminals monthly, or if there is a problem starting the*

## Cleaning the Fuel Filter Sediment Bowl

Before removing the glass or plastic sediment bowl, (often held in place by a wire hinge with a thumbscrew on the bottom) locate the fuel shut-off valve, which should be in the fuel line near the filter bowl. Turn it off. Then loosen the screw, swing the wire holder off to one side, and remove the bowl. Dump the contents into a container for safe disposal; wipe the bowl with a clean, lint-free cloth; and replace it. Turn on the fuel valve before tightening the lock nut completely so that the air in the bowl can escape as the bowl fills with fuel. When the bowl is full, finish tightening the nut.

Clean air filters regularly by blowing compressed air from the inside out.

*engine.* A battery's life is over when it can no longer hold a proper charge. A properly maintained starter battery should last at least five years, three in a hot climate. Sulfation is the leading cause of battery failure because it lessens a battery's ability to recharge. It occurs when hard lead sulfate crystals form on the lead plates of a battery. The two main causes of sulfation are not keeping the battery fully charged and letting electrolyte level fall below the top of the plates.

## TRUCK MAINTENANCE

Follow the heavy-duty service routine in your vehicle owner's manual. Routine servicing of the engine, cooling systems, suspension, tires, wheel bearings, brakes, and other mechanical components not only prolongs the life of a vehicle, but also allows you or your mechanic to uncover problems before they become emergencies. Common problems encountered in a hauling vehicle are overheated engines and transmissions, flat tires, and brake or hitch failures.

Be aware of how driving conditions affect wear and tear on your vehicles. For example, when towing or driving on gravel roads, a vehicle requires more frequent air filter and oil changes. When towing at night in certain parts of the country, bugs can plug the radiator and cause the engine to overheat. If you find that you need to clean bugs off your windshield, you can be sure that your radiator needs to be cleaned as well.

If your truck is serviced regularly, it will be ready at a moment's notice if you need to take your horse to the veterinarian or evacuate in case of emergency. For longer trips, follow the checklist in appendix B each time you use your truck.

## TRAILER MAINTENANCE

Just because a truck and trailer operate fine for short, local trips, does not mean the rig is roadworthy for longer hauls. In addition to keeping your truck in good shape, take good care of your trailer so that it is always ready to roll. Be sure to register and get new license plates or stickers on time each year so your trailer is ready to use when you need it.

*Clean the wheel bearings and then repack with grease annually or every 3000 miles.* Make sure that the seals, which keep out dirt and moisture, are replaced at the same time.

*Inspect the brakes on a new trailer after 200 miles of travel and then plan to service or adjust them every 3000 miles or according to the instructions in your trailer manual.* For a thorough brake check, the wheels and brake drums should be removed. Have the pads checked for wear, and replace if necessary.

If the brake pads have worn too thin, the rivet heads may have scored the drums — that's often the squeal you hear when the brakes are applied. (The noise could also be from dust accumulation.) A brake shop or your trailer dealer can test the brake magnets with an ohmmeter to see that they are drawing the proper number of ohms according to the specifications in your trailer owner's manual. If you have hydraulic brakes, make sure all fluid lines are in good condition and are not leaking.

*Rotate and balance tires once a year or every 5000-7000 miles to equalize wear.* Check for bare patches, bulges, and other defects. If wear is uneven, check axle alignment. Keep all tires inflated to proper level. Many tires lose 5–10 pounds pressure during winter storage.

*At least once a year, or every 3000 to 4000 miles, grease the springs and shackles, if your trailer has them.* Trailers with rubber torsion suspension don't have springs and don't require this maintenance. If your trailer has shock absorbers, check them and replace when necessary. Twice a year check the bushings where the shackles are pinned to the spring ends and the frame and also the bolts that secure the axle to the springs.

*Check the floorboards frequently for rotting, splintering, shrinking, or warping.* Replace boards that are even slightly damaged with pressure-treated lumber that is clear (no knots) and the same dimension as the existing floor. Treating the existing floor with a preservative can combat the effects of manure and urine. Use good-quality mats with "life" to help absorb road vibrations and shock, and replace them after they become excessively worn. (Old conveyor belt material makes terrible mats, because it has no cushion and is very slippery when wet.) Regularly check the plywood under the mats on ramps for signs of rot.

*Keep an eye on the interior of your trailer.* The bottom two feet of the sidewalls can sustain a lot of abuse from the hooves of scrambling horses. If the walls are metal, check for rust. You may wish to install thick mats or ¾-inch plywood over the metal

## Testing Your Brakes

Perform a brake check several times a year, with both an empty trailer and with a loaded trailer. A dry, hard, level roadway is ideal for this test. Accelerate to 10 mph and then brake. It helps to have a knowledgeable observer on the ground to tell you if a particular wheel is either locking up or rolling free in relation to the others. If necessary, the noncompliant brake should be adjusted. This may involve crawling under the trailer and using a screwdriver-type tool to adjust a tensioning screw within the brake drum. If you are not experienced with this, have your trailer dealer or a mechanic make the adjustment for you.

The next step is the brake controller test and adjustment. Refer to your brake controller manual or brochure for specific instructions. On level ground, accelerate to about 30 miles per hour. Then without using the brake pedal, bring the rig to a stop by using the manual electric brake controller mounted on or under the dashboard. If the trailer brakes grab suddenly with a lurch or can't stop the rig, adjust the controller per manual instructions.

Once the trailer brakes alone stop the rig properly, test by using your brake pedal, which activates your towing vehicle brakes and your trailer brakes simultaneously. The trailer brakes should not lock up on dry pavement when using the brake pedal.

walls for added protection for both the horse and your trailer. Regularly check for sharp edges, loose bolts, or protrusions that might injure a horse.

*Wash the trailer as needed and wax it at least twice per year.* Depending on the climate and usage, steel trailers usually need to be repainted every three to eight years. Sandblasting may be required to remove rust before painting. Aluminum trailers need an annual acid bath to keep them looking new. Check your yellow pages for commercial truck washes that offer this service to semi tractor/trailers.

*Every time you use your rig, check that the hitch, safety chains, chest bars, tail bars, dividers, doors,*

## Emergency Trailer Brake Check

One way to test the system is to remove the breakaway pin from the switch. This activates the electromagnets and engages the brakes — you might actually hear a "clunk." Then drive forward and see if your trailer brakes lock up. If they do, the battery and mechanism are operational. If the trailer were to become separated from the truck, the breakaway pin would pull out of the switch and the brakes would be engaged.

If your trailer brakes don't lock up at all during the test, either your battery needs charging or there is a problem in the wiring or the brake mechanism.

Since this test does not tell you if your battery is fully charged, you'll need to meter it to be sure. To guarantee that the emergency trailer brake battery is fully charged and operational, test it with a voltmeter. It should read 12 volts.

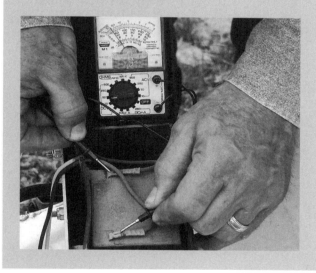

and windows work properly. Check safety chains for worn links or cracked welds. If you live in a humid climate, you'll need to clean and oil the gears of the hitch jack so that it moves up and down easily. Grease or soap the ball frequently to keep it moving freely in the coupler. Lubricate the moving parts of the coupler as necessary.

*Use spray lubricant on any hinges, latches, or other moving parts that do not function freely.* When they start to squeak or bind up, it is past time to give them a treatment. How often you need to lubricate hinges depends on the climate; it might be once a month or once a year.

*Check rubber gaskets and moldings around windows and doors to be sure they make a complete seal against rain and weather.* Replace when necessary. When replacing seals, you might need to first remove rust from the area with a putty knife and wire brush, and then apply a rust treatment and paint.

*Clean your rig after each trip so that it will take less time to get ready the next time.* Clean old feed out of the mangers and sweep the floors, hanging up the mats to let the stalls dry out. Hose the mats off when needed. Leave the tack room tidy.

# Storage

One of the most important recommendations for proper care and maintenance of farm equipment is to store it inside, out of the harmful effects of the weather. Sun, rain, and snow each bring their own brand of deterioration to machinery. Keeping tractors and implements indoors extends their useful life by protecting them from rust, ultraviolet deterioration, and damage from environmental moisture. It is also much nicer to hop on a dry tractor inside a shed rather than a soaking wet, snow-covered, or searing hot tractor that is stored outdoors.

Most compact tractors measure 7 feet, 3 inches at the top of the roll bar. Many new compacts offer a fold-down roll bar that breaks in the center to accommodate standard garage door heights of seven feet. Most utility tractors measure between 8 feet, 2 inches and 8 feet, 6 inches to the top of the roll bar or factory cabs. Farm tractors can be as tall as 12 feet.

The best way to determine storage requirements for your equipment is to measure each vehicle or implement, leaving enough space on all sides so that you can safely park something else alongside and so that you can safely move between pieces of equipment. The length times the width will give you the minimum square feet of storage space required.

Store your trailer on level ground with the hitch jack adjusted so that the trailer's weight is balanced between the tongue and the tires. If possible, park the trailer out of the weather to preserve the paint job, or buy a tight-fitting trailer cover. The rubbing of a loose cover might do more damage than good to the paint job. If outside, elevate the front end just a bit so that rain and snow slide off the roof.

Parking your trailer on concrete helps protect against tire rot. Most trailer tires usually have to be replaced due to deterioration from exposure to ultraviolet light and moisture before they are worn out from travel miles. If you are going to store a trailer for an extended period of time, jack up each side and place blocks under the axles where the springs attach to take weight off the tires.

Be sure the trailer is never parked or stored within reach of horses, as they will likely damage it by chewing or rubbing. If you want to park a trailer in a pasture to accustom young horses to it, use an old, safe trailer.

## Approximate Square Feet Required to Store Various Implements

| Implement | Square Feet |
|---|---|
| Tractor, garden | 40 |
| Tractor, compact | 60 |
| Tractor, utility | 80 |
| Tractor, farm | more than 80 |
| ATV/UTV | 15–60 |
| Disc (14 feet wide) | 140 |
| English harrow (14 feet wide) | 100 |
| Rotary harrow (8 feet diameter) | 70 |
| Mower (6 feet wide x 7 feet long) | 50 |
| Manure spreader (14 feet x 8 feet) | 130 |
| Blade | 25 |
| Posthole digger (standing) | 10 |
| Posthole digger (laying) | 20 |
| Flatbed trailer | 160 |
| Baler, small | 130 |
| Cart | 20 |

# Appendix A: Driving Tips

## SAFE DRIVING

Driving a large rig can be intimidating at first, but the best way to become a better trailer driver is to practice. If you have a level field at home, set up a course of cones or plastic barrels or buckets to maneuver around. When you feel confident with your skills at home, take your empty trailer to a parking lot to practice. During the week, church parking lots can provide open level "playing fields" for your trucker school drills. You might also try the local high school parking lot on weekends or a big department store lot during off-hours.

Eventually, you'll want to put a little pressure on yourself and practice maneuvering between cones in a quiet section of a parking lot during business hours. You'll be glad you did the first time you have to park your trailer in a tight space at a horse show or a vet clinic. It is far better to crunch a few cones in the grocery store parking lot than it is to smash into someone else's trailer at a show.

If you are new to trailering and are hesitant to take on the responsibility, don't worry. Confidence comes with experience. Before you start driving your own rig, go along with a competent seasoned hauler a few times to get a feel for the whole thing. Whether you are taking a practice trip or hitting the road for real, always bring someone with you, making sure it is a person you can rely on to assist you in case of emergency. Sometimes having the wrong person along can fluster you and undermine your confidence. The right traveling companion gives you a great feeling of teamwork and confidence, which usually means a safer trip.

## Using Your Mirrors

While you are driving, learn to use your side view mirrors to check behind you for approaching traffic or trailer doors that might have come unlatched. Since you will often be traveling at a slower speed than passenger cars, you should always be on the alert for cars coming up behind you to pass. Adjust your mirrors so you can see alongside and behind your trailer. When passing, use your mirrors to be sure that your trailer is well clear of the vehicle you passed before you start to pull back in front of it.

Have your spouse or a friend help you practice. When an assistant is helping you to back up or maneuver your trailer, it is his responsibility to be sure he can see you and you can see him in your side mirror at all times. Ask your helper to pay special attention to this, because it is very difficult for the driver to safely navigate while turning around

**Operating a truck and large rig takes some practice, so before you head out on a long trip, do some turning and backing in an empty parking lot.**

**Trailer mirrors extend out to the sides more than normal truck mirrors to give the driver better peripheral vision.**

**When backing into a tight spot, having someone guide you can be helpful.**

to look for a helper. Also, the signal for *stop* is pretty universal, but those for *turn left, turn right, a little bit more,* or *a lot more* can all look quite similar to the driver. Make some specific signal arrangements ahead of time and avoid a crunch or divorce court!

### Backing up a Trailer

Are you "backlexic" when it comes to maneuvering your trailer in reverse? At first, backing up can be very confusing, but with experience it will become second nature. The most important thing is to take your time. There is never any hurry when you are driving a rig and especially not when you are backing up. Don't hesitate to get out of the truck and walk around to see where you are trying to place your trailer. That often gives you a visual goal and allows you to see how much space you have and what obstacles might be in the way. If the mirrors are really confusing you, it's perfectly fine to turn around and look out the window.

When you are backing up a trailer, keep the following in mind: If you want the trailer to go to the left, the rear end of your truck must go to the right. That means you turn the steering wheel to the right (clockwise). You do this while looking in a mirror, which makes your brain do another left-right switch. If you are still having trouble knowing which way to turn the steering wheel, use the following tips.

- Put your hand in the center of the bottom of the steering wheel. If you want the back of the

trailer to move to the left, move your hand to the left.
- If you want the back of the trailer to go to the right, move your hand to the right.
- For a sharp turn, turn the steering wheel before you press the accelerator.
- For a gradual turn, turn the steering wheel and press the accelerator at the same time.
- Once the trailer is going in the direction you want, you need to straighten out the truck wheels to have your truck follow the trailer.

On your first trip to the parking lot practice area, set up a situation that requires you to back straight into an alley of cones or use the painted lines on the parking lot as guidelines.

Then set out a single cone and use it to become familiar with the turning radius of your rig as it moves forward. First turn to the left around the cone, keeping a close eye on it. Remember, the longer your rig, the more the rear wheels of your trailer will cut inside the track of your truck. Because goosenecks are hinged so far forward on the truck, the trailer tends to "cut the corner" and so requires a wider turn than a straight-pull.

Then use the cone to practice turning right. This is harder, as the cone is more difficult to keep track of in the right mirror. If you lose sight of the cone, it might be mashed the next time you see it!

Return to the backing exercises, this time incorporating turning along with the backing. Whenever possible, set up the situation so that you are backing with the bend in your rig on the driver's side where you can better see what is happening. However, eventually you will have to learn to back from the right where you are backing the trailer into a blind spot. Start out giving yourself plenty of room and make a large sweeping turn backward until you get lined up. Alternately looking in your mirror and out the back window might help or it might confuse you — experiment with what feels comfortable.

If you find that you turned too late and have overshot your target, it is better to pull far enough ahead to straighten out completely and try again. When making an adjustment, don't overcompensate. If you begin to jackknife (making such a sharp angle that your trailer might crunch the back corner of your truck), stop immediately and try again.

## Swaying or Fishtailing

It's likely that at some point during a trip you'll feel the trailer swaying from side to side. Swaying is more common with a straight-pull trailer than with a gooseneck. Swaying can be caused by many things: a flat tire, low air pressure on truck or trailer tires, an unbalanced load (usually because the trailer is too heavy towards the rear), a too-heavy trailer, too much speed (especially downhill), moving horses, wind gusts or turbulence from passing trucks, irregular or washboard road, and poor wheel alignment or balance.

Gentle swaying can usually be controlled by simply keeping the tow vehicle straight, letting up on the accelerator, and gradually slowing down. If your trailer sways regularly, stop and check the tires and make sure your horses aren't scrambling around. Rebalance your load if necessary by moving items from a rear tack room, for instance, to a front compartment.

Violent swaying or "fishtailing" can be very frightening and unnerving, but don't panic. Keep your eyes on the road, don't try to correct with the steering wheel, and *do not hit the brakes!* If you slam on the brakes, the trailer could jackknife and cause a serious wreck. Keep calm, continue straight ahead, and gently apply the manual trailer brake mounted under your dashboard while gradually letting up on the accelerator. This will diminish the swaying and allow you to maintain control of the rig. It is a good idea to practice this exercise ahead of time with an empty trailer until you feel you have overridden your reflex to put your foot on the brake pedal.

When you do apply the brakes, do so gradually or with a slight pumping motion to pulse the brakes without locking them. Never slam on the brakes unless an emergency situation leaves you no alternative, because the trailer is likely to jackknife and crash into the truck.

Slow speed limits on curves are there for a reason. Be considerate of your equine passengers and take curves at a safe speed.

When traveling in hilly or mountainous country, use a lower gear instead of constantly braking, which could cause your brakes to heat up and even catch on fire.

## ON THE ROAD WITH A TRAILER

- Never go over the speed limit, and don't be reluctant to drive under the speed limit. Let people pass you.
- Use turn signals well in advance. Don't make any quick lane switches.
- Realize that when pulling a trailer, you need a longer distance to stop your rig. Anticipate stop signs and corners, so you have ample distance for slowing or stopping.
- Slow down for obvious bumps or holes. Make all acceleration and deceleration gradual for your horse's comfort.
- Take curves at moderate speed to prevent scrambling and swaying. Some curve signs have speed limits posted, but don't wait for a sign to tell you to slow down when you will be twisting up or down a road. When taking a corner, wait until your vehicle is completely straight before you accelerate, otherwise your horses may lose their balance.
- Keep your senses alert to unusual sounds, smells, and vehicle motions. Keep your radio and CD player off so you can hear unusual sounds. Check your mirrors often for open doors, signs of smoke, or other unusual events.
- Use lower gears when going down hills. In a lower gear, your engine helps slow your vehicle and takes the load off the brakes, to help prevent brakes from overheating. Some hills have a low gear reminder, but many don't.
- Always take the responsibility to drive according to your load and the contours of the road.

# Appendix B: Trip Checklists

## TRAILER HITCH CHECK

Your trailer hitch and truck's receiver are essential for the safety of you and your horse. Develop a method of hooking up your trailer and do things in that order every time. After you have finished the process, go over each aspect again. Before you drive off, check everything related to the hitch, receiver, breakaway cable, and electric brakes once more.

Be sure your hitch is not cracked or rusted, that it does not have loose parts, and that the trailer coupler is securely seated over the ball with the locking mechanism engaged. Whenever you plug or unplug the electrical wire, do it by holding onto the metal plug not the wire sheath. It won't take more than a few times yanking a plug out by the wire for you to start having electrical problems.

When you get ready to move your trailer, raise the hitch jack so the wheel is at its full height. If you don't jack the wheel all the way up, when you go over uneven ground or even just in or out of a driveway, the wheel could drag on the ground. If it hits hard enough, it could pop the hitch loose.

Some trailer manufacturers make the hitch wheel removable. Be sure to put the pin clip back in the wheel and stow it somewhere safe so you can find it when you get to your destination.

## HOOKING UP YOUR TRAILER

When you are first breaking in your coupler, it might be difficult to slip it onto the ball. You can help it glide into place by rubbing grease or a dry bar of soap over the ball. (Soap is not as messy as grease, and this is a good way to use up soap scraps.) A well-lubricated ball allows the coupler to move freely over it and reduces ball and coupler wear from friction. Make sure the nut securing the ball to the ball mount is tight. With a fifth-wheel, lubricate the head plate or use a good nylon lube disc.

Once you get the coupler resting on the ball, you might need to give the coupler a kick to close the hinged portion of the coupler so the collar of the hitch slides over it.

## Trailer Checklist for Trips

- ❏ Tire pressure (inflated to the specification in your owner's manuals?)
- ❏ Wheel lug nuts (tight per trailer specs?)
- ❏ Tires (irregular tread wear, bulges, defects, weather checking?)
- ❏ Spare tire and jack (in place and functional?)
- ❏ Hinges, latches, or other moving parts (need oil?)
- ❏ Running lights, turn signals, brake lights, emergency flashers (working?)
- ❏ Brakes (working?)
- ❏ Inside of the trailer (clean and free of hornet or mouse nests?)
- ❏ Bedding (needed for this trip?)
- ❏ Hitch and safety chains (in place?)
- ❏ Trailer registration and other paperwork (in the towing vehicle?)

## Truck Checklist for Trips

- ❏ Tire pressure (inflated to the specification in your owner's manuals?)
- ❏ Gas tank (full?)
- ❏ Oil (level okay?)
- ❏ Power steering fluid (level okay?)
- ❏ Transmission fluid (level okay?)
- ❏ Brake fluid (level okay?)
- ❏ Coolant (in both radiator and recovery reservoir?)
- ❏ Windshield washer fluid (topped off?)
- ❏ Wiper blades (functional?)
- ❏ Air filter (clean?)
- ❏ Battery terminals (clean of corrosion?)

Know the proper tire pressure for your truck and trailer tires and check them before a trip and while traveling.

## Breakaway Mechanism

To fasten the cable of the breakaway mechanism to your truck, first remove the breakaway pin from the switch. This requires a strong pull. Wrap the wire cable around your gloved left hand and grab the cable close to the coupler with your right hand. Give a sharp jerk straight backward to remove the pin from the switch.

Run the cable around the frame hitch on your truck and pass the pin through the loop on the other end of the cable. Pull the pin end of the cable to take the slack out of the cable.

Plug the pin back into the switch.

Remember that the voltage of the breakaway battery should be checked several times a year and trickle charged when necessary.

## Tire Pressure

Check the tire pressure on your truck and trailer before every trip when the tires are cold. It is normal for a warm tire to run with a slightly higher air pressure. Look for the maximum specification, such as 50 psi, on the sidewall of your tires. That indicates the maximum tire pressure. Be sure to not inflate the tires over that pressure. Refer to your owner's manual for the optimum tire pressure.

If you let your tires get low, they could overheat and cause blowouts. You should routinely check both your truck and trailer tires for irregular tread wear, bulges, defects, or weather checking so just before or during a trip you don't find any surprises that cause you unnecessary delays.

Check the tire pressure on the trailer spare. Be sure your spare is located in a safe place where it cannot be stolen. If it is on the outside of the vehicle, be sure it has a lock on it and know exactly where the key to the lock is located.

Check the tire pressure on the truck spare. This may require you to crawl under your vehicle. You'll see once you get under your truck to take this reading why it is important to stow the spare so that the stem side of the wheel is facing the ground.

## Wheel Lugs

Check to see that wheel lug nuts are tight. This is emphasized by many trailer manufacturers in owners' manuals and via decals often strategically placed on the trailer fender. You need to use a torque wrench to be certain you are meeting the manufacturer's specifications.

## BE PREPARED FOR FLAT TIRES

Learn where the various portions of your jack and crank are located. Often they are stowed in separate locations. Take the jack out and practice using it before you are caught in an emergency. Know where the spare tire is and know how to retrieve it and refasten its carrier. If your trailer has hubcaps, you should carry a tool to remove them. A four-way (star) wrench will likely have sockets to fit both your truck and trailer lugs. The star design makes it easy to apply a lot of balanced force to loosen and tighten wheel lugs. Use a dab of nail polish or paint to designate which arm of the star wrench has the socket to fit which lugs.

Have a tongue wheel chock, a jack, flat boards to support a jack in soft ground, blocks to block the wheels if you need to unhitch, and a special ramp that can replace the jack when changing tandem trailer wheels. Some ramps are suitable for trailers up to 20,000 pounds. Some are specifically designed for horse trailers with torsion bars or leaf or slipper springs. Buy one that is specifically made for the size and type of trailer you have.

Trailer wheel lugs should be tightened to manufacturer's specifications with a torque wrench.

## Emergency Kit for Truck

**For a flat tire:** Carry a tongue-wheel chock, a jack, flat boards to support the jack in soft ground, blocks to block wheels if you need to unhitch, and a ramp for changing tandem trailer wheels.

**For a disabled vehicle:** Have a tow chain on board to move your rig off the road or help pull it out of soft ground, and jumper cables to boost a dead battery.

Keep a prepackaged tool kit with a set of open-end wrenches, sockets, screwdrivers, and pliers along with a safety knife and other items.

Include a can of WD-40 lubricant, a pocket-sized multi-tool, additional pliers, vise-grips, and a hammer.

Carry three collapsible, reflective, weighted triangles to alert oncoming vehicles that you are stopped ahead. Have flares on board and know how to use them in an emergency.

**Quick fixes for common problems**: Make up a box with extra motor oil, a radiator hose, a fan belt, fuses, and bulbs. Hoses and belts are commonly replaced before they give out, so ask your serviceman to save the functional old ones for your emergency kit.

Electrical tape for wire repairs can also be used to keep trailer parts from rattling and for additional bandage material.

Extra bungee cords come in handy for holding open doors on windy days and securing broken parts such as dividers and grills, and for securing objects in the tack room, on the roof rack, or in the pickup bed.

Rags, paper towels, gloves, a small tarp to lie on, and old shirt to protect your clothes can make it easier to make repairs.

Buy an 11-ounce fire extinguisher containing potassium bicarbonate that is rated for grease, gasoline, and electrical fires and keep it charged and accessible.

A large waterproof flashlight or lamp will help if you need to change a flat tire in the dark. Battery-powered headlamps (hands free) are very useful. Always carry extra, fresh batteries, because night breakdowns often last longer than the set you have in your flashlight.

# Glossary

*2WD.* Two-wheel drive; see 4 × 2.

*4 × 2.* A four-wheel vehicle having power delivered to two wheels, usually the rear.

*4 × 4.* A four-wheel vehicle having power delivered to all four wheels.

*4WD.* Four-wheel drive; see 4 × 4.

*6 × 2.* A six-wheel vehicle (three axles) having power delivered to only two wheels, usually the front pair of tandem rear wheels.

*6 × 4.* A six-wheel vehicle (three axles) having power delivered to four wheels, usually both tandem rear axles.

*6 × 6.* A six-wheel vehicle (three axles) having power delivered to all six wheels.

*A-frame.* The front portion of a straight-pull trailer that extends ahead of the box.

*aerator.* A drum with spikes that is pulled behind a tractor to aerate compacted soil.

*alternator.* A type of generator used with internal combustion engines; it charges the battery and powers all electric systems when the engine is running.

*APV (all-purpose vehicle).* Includes all-terrain vehicles and utility task vehicles.

*ATV (all-terrain vehicle or four-wheeler).* A category of small, open, motorized vehicles, usually gasoline powered, with a straddle seat and handle bar steering.

*auction.* A public sale to the highest bidder.

*auger.* See *posthole digger.*

*automatic transmission.* A transmission that shifts gears automatically without the use of a clutch and gear shifter.

*AWD.* All-wheel drive.

*axle.* The shaft that connects a pair of wheels; typically there are two axles per vehicle.

*backhoe.* Excavation machine that digs by pulling a boom-mounted bucket toward itself; used mainly to dig footings and trenches.

*bale spears (tines).* An addition to the front or rear of a vehicle such as a tractor that enables you to spear large bales to move them.

*baler.* Either a self-contained piece of machinery or an implement pulled behind a tractor that gathers cut plant materials out of windrows and compresses it into round or rectangular (square) bales that are tied with twine or wire to make feed or bedding.

*ball.* The trailer ball, located on the towing vehicle, attaches to the trailer coupler to connect the trailer to the tow vehicle. Balls vary in size, capacity, and finish.

*ball height.* Measurement from the ground to the top of the ball on the truck or to the top of the inside of the coupling on the trailer when parked on a flat surface and parallel to the ground; the difference between the two determines the type of ball mount needed to make the trailer ride parallel to the ground when being towed.

*ball mount (also ball platform, shaft, shank, stinger, or drawbar).* The part of a receiver-style hitch that slides into the permanently attached portion of the hitch and is secured with a pin and clip; the ball is bolted onto the ball mount. Available in load-carrying and weight-distributing configurations. An adjustable ball mount allows a hitch ball to be raised and lowered in increments to enable the trailer to ride level.

*BCW (base curb weight).* The weight of a vehicle with standard equipment and a full tank of fuel. It does not include passengers, cargo, or optional equipment.

*blade.* A cutting edge and moldboard used for moving or rearranging dirt, manure, bedding, or snow; usually mounted on the rear or middle of the tractor.

**block.** The major casting of the engine; holds the cylinders, crankshaft, and camshaft (through the 1960s). When you look at an engine, most of what you see is the block.

**book value.** Originally referred to the Kelley Blue Book, dealers use this industry guide to estimate wholesale and retail vehicle pricing. The term now refers to a price looked up in one of the many pricing guides available online and in the reference sections of public libraries.

**brake controller.** A control unit mounted inside the tow vehicle that activates the electric trailer brakes in harmony with the braking of the tow vehicle; can also be used to adjust trailer brake intensity or to manually activate the trailer brakes.

**breakaway cable.** A safety device that connects the electric trailer brake's breakaway switch to the towing vehicle; if the trailer becomes disconnected from the hitch, the cable is pulled away and the switch activates the brakes.

**breakaway switch.** A safety device that activates the trailer brakes in the event the trailer accidentally becomes disconnected from the hitch while traveling.

**brush hog.** A type of rotary mower for rough-cutting and shredding heavy grasses or brush.

**bumper.** A wide bar, usually metal or rubber, attached to either end of a vehicle to absorb impact in a collision and protect the vehicle from damage.

**bumper-pull.** A misnomer; see straight-pull and tag-along.

**calcium chloride.** A compound mixed with water to fill tractor tires for ballast; does not freeze, but can cause the wheels to rust.

**cargo weight (CW).** All the weight added to the base curb weight (BCW), including the passengers, the cargo, and any optional equipment; when towing, the trailer tongue weight is also included in the cargo weight.

**category.** A rating from 0 to 3 that refers to the size and capacity of tractor hitches and implements.

**chassis.** See *frame.*

**checking (weather checking, weather cracking, or ozone cracking).** A pattern of small cracks that occurs on the surface of all tires as the rubber loses elasticity and begins to deteriorate; accelerated by exposure to sunlight and weather. Cracks may be cosmetic in nature if they don't extend past the rubber's outer surface or may be a reason to replace the tire if they reach deep into the rubber.

**class.** A rating from I to V that refers to the size and capacity of a trailer hitch; each class has a maximum tongue weight (TW) limit and a maximum gross trailer weight (GTW) limit. See *hitch ratings.*

**clip.** Used to retain the pin in a receiver-style hitch.

**clutch.** A device for connecting and disconnecting the engine and the driveshaft.

**continuous horsepower.** The engine power available for continuous-duty conditions within a specified speed range.

**coupler.** The trailer portion of a hitch that attaches the trailer to the tow vehicle by fitting over and locking to the hitch ball.

**creeper gear.** A very low gear used to pull implements that must move more slowly than standard first gear would allow.

**curb weight (CW).** The weight of a vehicle as it sits on the lot, including optional equipment but not weight of cargo or passengers.

**custom hitch.** One that is made to fit a specific vehicle.

**CW.** See *cargo weight* and *curb weight.*

**diesel engine.** A diesel engine ignites fuel in the cylinder from the heat generated by compression; the fuel is an oil instead of gasoline and no spark plug or carburetor is required.

**differential.** A set of mechanical gears that transfers the power of the rotating driveshaft to the rear axle and allows the wheels to turn at different speeds. The differential prevents the wheels from binding when turning corners as the outside wheel travels farther and turns more quickly than the inside wheel.

**disc harrow.** A harrow that uses one or more rows of adjustable discs to break up and smooth soil.

**disc plow.** A plow that uses rows of discs to turn over soil.

**distributor.** A component of an electric ignition system that uses a rotor to fire each spark plug wire in turn; replaced in later engine designs by electronic ignition.

**drag harrow (also flex-tine, finger, tine-tooth, weeder, or spike-tooth harrow).** A harrow with horizontal rigid bars that hold pointed, adjustable metal rods to dig into the soil.

**drawbar (also hitch bar).** A bar attached to the rear of a tractor for towing or dragging; also refers to a receiver-style hitch or fixed-hitch style ball mount. Also, a long bar across the front of an English harrow.

**drawbar horsepower.** Tractor horsepower that is measured at the tractor's drawbar and includes the power required to move the tractor.

**drill.** A wheeled implement that drops grains into soil openings created by a disc, a shovel, or other means.

**driveshaft.** A rotating shaft that transfers power from the engine to another component, typically to the differential.

**drop-down.** A ball mount that steps down to lower ball height to make the trailer ride parallel to the ground.

**dually.** A pickup truck, or light-duty tow vehicle, with four tires on one rear axle.

**duallies.** Dual tires.

**DW (dry weight; also unloaded vehicle weight).** The weight of a vehicle without added fuel, water, propane, supplies, and passengers, or any dealer-installed options.

**engine oil cooler.** A heat exchanger, similar to a small radiator, through which engine oil passes and is cooled by airflow.

**English harrow (chain, blanket, flexible tine harrow or harrow-rake).** A harrow made of many identical interlocking pieces of bent steel rod often with downward protrusions; looks like a very heavy chain-link fence.

**EROPS (enclosed rollover protective structure).** Adds weather-tight enclosures including doors and windows to a ROPS (rollover protective structure), allowing the operator to be totally enclosed.

**fifth-wheel.** A type of hitch that mounts over the rear axle in the bed of a pickup truck similar to hook-ups on big tractor/trailer rigs.

**fifth-wheel trailers.** Trailers designed to be coupled to a fifth-wheel hitch; can have one, two, or three axles and are the largest type of trailer built.

**fishtailing.** See *sway*.

**fixed tongue hitch (also fixed drawbar hitch).** A permanent hitch that remains in place when the trailer is unhitched. This type of hitch can be used as a weight-carrying hitch only.

**FOPS (falling object protective structure).** Typically a mesh sheeting structure attached to a tractor or other vehicle to protect the operator from branches, rocks, hay bales, and other falling objects.

**forklift.** A vehicle equipped on the front with a pair of flat tines or forks used for lifting and moving heavy loads, most often on pallets.

**frame.** Chassis; the part of a vehicle to which all other parts are attached.

**frame hitch.** A straight-pull hitch (as opposed to a gooseneck or fifth-wheel hitch) designed to be bolted or welded to the vehicle frame.

**front-end loader (also loader).** A bucket loader attachment designed to lift materials; usually hydraulically operated and mounted on the front of a tractor.

**FWD.** Front-wheel drive.

**GAW (gross axle weight).** The total weight supported by each axle of a vehicle.

**GAWR (gross axle weight rating).** The manufacturer's rating for the maximum allowable weight that an axle is designed to carry; GAWR applies to tow vehicle, trailer, fifth-wheel, and motor home axles.

**GCVW (gross combined vehicle weight).** The gross vehicle weight of the tow vehicle plus the weight of any towed vehicles or trailers

**GCWR (gross combination weight rating).** The maximum allowable weight of the combination of tow vehicle and trailer; includes the weight of the vehicle, trailer, cargo, passengers, and a full load of fluids (water, propane, fuel, etc.).

**generator.** A device connected to and powered by an internal combustion engine to charge the battery and to power all electric systems when the engine is running.

**gooseneck.** A type of trailer with a bunk that extends over the back of the tow vehicle and a vertical coupler that extends down from the bottom of the bunk to connect to a gooseneck ball mounted in the bed of the tow vehicle; a gooseneck hitch uses a ball $2^5/_{16}$ inches or 3 inches in diameter.

**gray market tractor.** A tractor, usually a compact diesel, imported from another country, often Japan; they are not required to comply with current OSHA safety equipment regulations such as rollover protective structures; some are retrofitted with necessary safety features to comply with U.S. rules.

**GTW (gross trailer weight).** The actual weight of the fully loaded trailer plus the tongue weight.

**GVW (gross vehicle weight).** The actual weight of a vehicle when fully loaded; includes passengers, cargo, fluids, and hitch weight. For trailer see *LTW*.

**GVWR (gross vehicle weight rating).** The total allowable weight of a vehicle, including passengers, cargo, fluids, and hitch weight.

**ground clearance.** The minimum distance to the ground from the lowest part of the underside of a vehicle.

**harrow.** An implement that is used to break up and smooth plowed or clumped soil; types include disc harrow, drag harrow, spring-tooth harrow, spike-tooth harrow, English harrow, and rotary harrow.

**hay rake.** An implement that is used to rake up cut grass into windrows so that it can be baled.

**hitch bar.** See *drawbar*.

**hitch ratings.** Trailer hitches are rated according to the maximum amount of weight they are engineered to handle. Class I units are rated for towing up to 2000 pounds, Class II for loads up to 3500 pounds, Class III for up to 7500 pounds, and Class IV for loads of up to 10,000 pounds. Class I and II receiver-style hitches can only be used as weight-carrying hitches, while Class III and IV receiver-style hitches can be used either as weight-carrying hitches or as weight-distributing hitches when a weight distribution system is added. Class V hitches can tow loads up to 14,000 pounds. These ratings may vary depending on the manufacturer. Fifth-wheel ratings range to 25,000 pounds.

**hitch weight.** See *tongue weight*.

**horsepower.** A measurement of energy; one horsepower is the amount of energy required to lift 550 pounds one foot in one second.

**hydraulic assist transmission.** A transmission that uses a hydraulic clutch to enable operator to change speeds and direction without stopping.

**hydraulic pump.** A pump that circulates hydraulic fluid within a closed system to provide pressure to operate power steering, loaders, and other implements and systems.

**hydraulics.** Systems that use pressurized oil to provide power to raise and lower a three-point hitch and other attachments, and to power attached or towed implements that have hydraulic components; live hydraulics can maintain oil pressure even when the clutch is disengaged.

*hydrostatic transmission.* A transmission that has infinite speed selection, simple direction reversal, and dynamic braking.

*implement.* A tool or device that does not run on its own power, but is pulled and/or powered by a tractor or other vehicle.

*jackknife.* When the tow vehicle and trailer fold at the hitch like a jackknife and collide with one another; can occur when turning too sharply, especially while backing, or when going downhill on a slippery surface and applying the tow vehicle brakes.

*joystick.* A sticklike controller, typically used on a hydraulic loader that is held in one hand and pivots at its base to control an implement's movements in several directions.

*limited slip.* A differential that limits rotation in the differential gear so that more power goes to the wheel with the most traction.

*live load.* Weight on a structure consisting of movable objects, animals, and persons.

*loader.* A hydraulically powered implement with a bucket that can pick up, move, and dump bulk materials such as dirt, gravel, manure, and bedding.

*loafing shed (also run-in shed).* A three-sided building, usually pole framed, that has the fourth side completely open or with one or more large openings.

*long bed.* A pickup bed that is eight feet long or longer.

*LTW (loaded trailer weight).* The weight of the trailer fully loaded minus the tongue weight.

*lug.* One of several bolts that attach a wheel to an axle. Also a deep, barlike tread on a tire.

*manure spreader.* A mechanized trailer that distributes a load of manure evenly across an area.

*MFWD.* Mechanical front-wheel drive.

*MLTW (maximum loaded trailer weight).* The maximum allowable fully loaded weight of a trailer; see *GCWR* and *GVW*.

*MTWR (maximum tow weight rating).* The maximum allowable towing capacity of a vehicle.

*moldboard.* The curved surface of a blade or plow.

*NFE (narrow front-end).* A tricycle front-end on a tractor with one or two wheels. Made in the 1950s; great maneuverability but unstable.

*OEM (original equipment manufacturer).* Parts and equipment that come on your vehicle from the factory.

*pin.* A short rod used to fasten a ball mount into a receiver hitch; the pin, or kingpin, projects from the pinbox plate on a fifth-wheel trailer to engage a hole in the head plate on the truck hitch, thus connecting the trailer to the truck.

*pitman arm.* A shaft used in combination with an off-set flywheel to convert circular motion to back-and-forth motion; commonly seen on sickle bar mowers, it is usually made of wood so that it will self-destruct if it comes in contact with rocks.

*plug.* The connector on the trailer end of the wiring connection that inserts into a socket on the truck to connect the trailer wiring to the tow vehicle's electric system.

*posthole digger.* An auger that digs a hole for a fence post; can be powered by a PTO, hydraulics, or an integral gas engine.

*power shift transmission.* A transmission that uses a hydraulic clutch connecting a gear to a shaft to change speed and direction when the tractor is in motion; there is no power loss during shifting as with a hydrostatic transmission.

*PTO (power take-off).* A system that uses the tractor's horsepower to turn an integral shaft that supplies power to implements connected to it by another detachable shaft. Common PTO operating speeds are 540 rpm and 1000 rpm, which are usually the standard running speeds of the tractor; live, or independent, PTOs rotate independently of the tractor drivetrain; semi-live PTOs can be engaged and disengaged in conjunction with the tractor drivetrain by operation of a two-stage clutch where you push the clutch in half-way to shift gears and all the way to stop the PTO.

**receiver.** The part of a straight-pull trailer hitch that attaches to the frame of the tow vehicle and has a square receptacle (1½-inch, 1⅝-inch, or 2-inch square) for inserting a ball mount or other attachment.

**receiver hitch (also box or tube hitch).** A hitch in which the ball mount (or ball platform) can be removed from the permanently attached portion of the hitch (the receiver) when not needed.

**ROPS. (Rollover protective structure).** A roll bar to help protect the operator in case the vehicle tips over. ROPS are standard equipment or optional on later-model U.S. tractors and are mandatory on tractors of a certain minimum weight. Called rollover protective system when used in conjunction with a seatbelt.

**rotary harrow.** A harrow consisting of a large horizontal wheel rimmed with downward-pointing spikes; typically connects to the three-point hitch of a tractor.

**rotary mower.** An enclosed spinning blade for mowing weeds and grasses, brush, and small trees; low profile is good for getting around rocks and under low-hanging trees.

**rpm (revolutions per minute).** Turning 360 degrees is considered one revolution for an engine crank shaft.

**RWD.** Rear-wheel drive.

**safety chains.** A set of chains that are attached to the trailer bunk or A-frame and connected to the tow vehicle while towing. Safety chains are intended to keep the trailer attached to the tow vehicle in the event of hitch failure, preventing the trailer from complete separation; with a straight-pull trailer they should be installed in an X-pattern (criss-crossed) so the coupler is held off the road if it becomes disconnected from the hitch.

**shank.** See *ball mount.*

**short bed.** A pickup bed that is shorter than eight feet, usually six and a half feet.

**shuttle-shift transmission.** A mechanical or hydraulic lever that shifts between forward and reverse while staying in the same gear.

**sickle bar mower.** A mower consisting of a long bar of cutting blades, half of which are pulled back and forth by a pitman arm, to cut grass crops on relatively smooth surfaces.

**single axle.** An axle that is not part of an axle group.

**skid steer (also skid loader).** A compact machine steered by means of two braking levers that control the speed of the wheels on the left and right sides. More maneuverable than a conventional tractor with similar load capacity, it is commonly used with a front loader attachment for pushing or lifting material and for digging, but can be fitted with a variety of specialized buckets or attachments.

**slider hitch.** A sliding hitch used on short-bed pickup trucks to enable them to tow fifth-wheel trailers; when driving at slow speeds the hitch slides toward the back of the bed on rails, automatically or manually, to allow sufficient clearance for tight turns without having the trailer hit the cab of the truck.

**slow-moving vehicle emblem.** A bright orange triangle symbol indicating a slow-moving vehicle; required on the rear of farm machinery being driven on public roadways.

**socket.** The wiring connector on a tow vehicle into which the wiring plug from the trailer is inserted.

**spike-tooth harrow.** A harrow that uses downward projecting teeth, rods, or spikes to break up and smooth the surface of the ground.

**spreader.** An implement that distributes seeds, fertilizers, manure, compost, and other loose materials quickly and evenly.

**spring bar.** The lever part of a weight-distributing hitch that does the lifting; typically made of spring steel.

**spring-tooth harrow.** A harrow that uses spring steel bars shaped in half-moons to break up and smooth soil.

**straight-pull trailer (also tag-along trailer).** A trailer that is pulled behind a vehicle with a frame hitch as opposed to a fifth-wheel or gooseneck.

**suburban tractor.** Lawn or garden tractor.

**suspension.** Components, including springs and shock absorbers, which secure the axle or axles to the frame of a vehicle.

**SUV.** Sport utility vehicle.

**sway.** Fishtailing or yaw; the dangerous swinging back and forth action of a trailer usually caused by the trailer load being imbalanced to the rear.

**sway bar (also anti-sway bar, stabilizer bar, or anti-roll bar).** A solid rod made of spring steel that mounts to the vehicle's frame with the ends connecting to the right and left suspension components. Can be used on the front and the rear to help reduce vehicle lean or roll during cornering and when towing; it is sometimes confused with a weight-distributing hitch, but it is not part of a trailer hitch.

**sway control.** One or more devices connected from the A-frame of a straight-pull trailer to the hitch-ball platform of the tow vehicle to dampen the swaying action of a trailer, either through a friction system or a "cam action" system that slows and absorbs the pivotal articulating action between tow vehicle and trailer.

**synchromesh.** See *synchronized transmission.*

**synchronized transmission.** A manual transmission that uses clutches to equalize the speeds of the gear and the shaft, allowing for smooth shifting of gears while moving.

**tag-along trailer.** See *straight-pull trailer.*

**tandem axle.** Two axles, generally more than 40 inches and less than 96 inches apart.

**three-point hitch.** A system at the rear of a tractor to which implements are mounted; the three primary parts of this hitch system are the top link and two lower lift arms; adjusting one or all of these components allows the tractor operator to raise, lower, or tilt an implement.

**tie rod.** The rod that connects the steering arms of the two front wheels.

**tine.** A long, pointed steel prong mounted on a tractor or other vehicle and used to pick up bales of hay; a tooth or spike of a harrow.

**tongue.** The forward-most portion of a trailer to which the coupler is mounted. See *coupler.*

**tongue weight.** See *TW.*

**torque converter.** A hydraulic torque multiplier in the drive train of a vehicle that acts as a fluid coupling; typically connected to the transmission.

**tow rating.** The manufacturer's rating of the maximum weight limit that can safely be towed by a particular vehicle; criteria for determining tow ratings include engine size, transmission, axle ratio, brakes, chassis, cooling systems, and other special equipment.

**trailer.** A non-self-powered towed vehicle used for carrying cargo.

**trailer brakes.** Brakes that are built into the trailer axle systems and are activated either by electric impulse or by a surge mechanism; electric trailer brakes can be activated simultaneously with the tow vehicle's brakes or manually by means of a brake controller. Surge brakes utilize a mechanism positioned at the coupler to detect when the tow vehicle is slowing or stopping and activate the trailer brakes via a mechanical or hydraulic system.

**transfer case.** A part of a four-wheel-drive system that connects to the transmission and also to the front and rear axles by means of driveshafts;

the transfer case receives power from the transmission and sends it to both the front and rear axles.

*transmission.* Gears and other elements assembled together to allow a variation in a vehicle's speed or direction.

*transmission cooler.* A heat exchanger similar to a small radiator through which automatic transmission fluid passes and is cooled by airflow.

*tricycle.* See *NFE.*

*turbo charger.* A device that compresses air and forces it into the combustion chamber of an engine to increase horsepower.

*TW (tongue weight; also hitch weight).* The amount of weight imposed on the hitch when the trailer is coupled. Tongue weight for a straight-pull trailer is typically 10 to 15 percent of total trailer weight; gooseneck and fifth-wheel tongue weight is approximately 25 percent of the total trailer weight.

*umbilical cord.* See *wiring harness.*

*utility tractor.* A tractor of approximately 45 to 85 horsepower.

*UTV (utility task vehicle).* A category of small, motorized carlike vehicles, powered by gasoline or electric engines, with a single or bench seat and steering wheel.

*UVW (unloaded vehicle weight).* Dry weight; the weight of a trailer without added cargo; the manufacturer's UVW will not include any dealer-installed options.

*VIN.* Vehicle identification number.

*weight-carrying.* A towing situation wherein all of the tongue weight is carried directly on the hitch.

*weight-carrying hitch (also dead-weight hitch).* A towing system that accepts the entire hitch weight of the trailer; in the strictest sense, even a weight-distributing hitch can act as a weight-carrying hitch if the spring bars are not installed and placed under tension.

*weight-distributing.* A towing situation whereby some of the tongue weight is distributed by use of a weight-distributing attachment to the tow vehicle's front axle.

*weight-distributing hitch (also equalizing hitch).* A straight-pull hitch system that utilizes two spring bars, one on each side of the trailer A-frame, to lift up and apply leverage to the tow vehicle to distribute a greater portion of the trailer's hitch weight to the tow vehicle's front axle and the trailer's axles, resulting in greater vehicle stability while towing.

*weights.* Metal or liquid weights added to a tractor or implement for balance, traction, stability, and/or digging force.

*WFE (wide front-end).* Tractors without tricycle front ends. All tractors since 1960 have been built with WFE.

*wheelbase.* The distance from the centerline of the vehicle's front axle to the centerline of the rear axle.

*wiring harness.* A bundled group of wires, usually color-coded, that have a plug or socket attached; typically used to connect a vehicle's electric system to that of a trailer or implement.

*zerk.* A tiny one-way fitting through which grease is injected into bearings and bushings.

# Resources

For state regulations regarding trailer brakes and chains and other trailer laws, contact your State Motor Vehicle Division or Office of Motor Carriers.

For federal regulations, contact the Department of Transportation: 400 7th Street, SW, Washington, D.C. 20590, 800-877-8339, www.dot.gov or the Federal Highway Administration: 400 7th Street, SW, Washington, DC 20590, 202-366-0660, www.fhwa.dot.gov

For ATV safe driver courses and information, contact the ATV Safety Institute: 2 Jenner Street, Suite 150, Irvine, CA 92618-3806, 800-887-2887, www.atvsafety.org, or the Canada Safety Council: 1020 Thomas Spratt Place, Ottowa, ON K1G 5L5, 613-739-1535, www.safety-council.org/training/ATV/atv.htm.

Many local retail stores carry farm equipment. Check Tractor Supply, Sears, Home Depot, and Ace Hardware, for example. For specific types of equipment, try the following companies.

## ATVS, GOLF CARS, LAWN TRACTORS, AND ACCESSORIES

**Agri-Fab**
809 South Hamilton Street
Sullivan, IL 61951
217-728-8388
www.agri-fab.com
*(carts, ball hitch kits, sprayers, electric spreaders, steel tow rollers, trail mowers, three-point adapter trailers, chippers)*

**American Honda Motor Company**
Power Equipment Division
4900 Marconi Drive
Alpharetta, GA 30005-8847
678-339-2600
www.honda.com

**Arctic Cat Inc.**
601 Brooks Avenue South
Thief River Falls, MN 56701
218-681-8558
www.arcticcat.com

**AUSA Corp.**
2655 Le Jeune Road, Ste. 916
Coral Gables, FL 33134
800-820-2872
www.ausa.com

**Bobcat**
P.O. Box 6000
West Fargo, ND 58078-6000
(800) 743-4340
www.bobcat.com

**Bombardier**
726 Saint-Joseph Street
Valcourt, Québec
Canada J0E 2L0
877-469-7433
www.bombardier-atv.com

**Bramco**
513 North Main
Fairview, OK 73737
580-227-2345
www.bramcoinc.com

**Club Car, Inc.**
P.O. Box 204658
Augusta, GA 30917-4658
800-258-2227
www.clubcar.com

**Country Mfg. Inc.**
P.O. Box 104, 333 Salem Avenue
Fredericktown, OH 43019
740-694-9926
www.countryatv.com
*(harrows, trailers, manure spreaders)*

**Cub Cadet**
P.O. Box 368023
Cleveland, OH 44136
800-422-3381
www.cubcadet.com

**D&P Industries, Inc.**
P.O. Box 1828, 1523 SW Juniper
Redmond, OR 97756
877-923-1992
www.thecomposter.com
*(electric chipper)*

**Diversified Golf Cars, Inc.**
5501 Commerce Drive, Suite 104
Orlando, FL 32839
407-851-9353
www.diversifiedgolfcars.com
*(utility bodies, cargo boxes, trailers)*

**DR Power Equipment**
127 Meigs Road
Vergennes, VT 05491
800-687-6575
www.drpower.com
*(lawn vacuums, mowers, chippers, trailers, sprayers, harrows, powered wagons)*

**Electric Car Distributors, Inc.**
71-441 Highway 111
Rancho Mirage, CA 92270
800-476-5642
www.electriccar.com

**E-Z-Go**
1451 Marvin Griffin Road
Augusta, GA 30906
800-241-5855
www.E-Z-GO.com
www.cushmanco.com

**Golf Car Utility Systems**
419 South 7th
Quincy, IL 62301
800-383-8845
www.golfcarus.com
*(utility bodies for golf cars, boxes, cargo beds, flat beds, dump beds, trailers, mirrors, seats, snowblades, windshields)*

**Gorilla Vehicles**
16121 St. Croix Circle
Huntington Beach, CA 92649-1715
www.gorillavehicles.com
714-377-7776
*(electric ATV)*

**Husqvarna**
7349 Statesville Road
Charlotte, NC 28269
800-487-5962
www.husqvarna.com

**Jacobsen Division of Textron**
One Bob-Cat Lane
Johnson Creek, WI 53038
920-699-2000
www.jacobsengolf.com

**Kawasaki Motors Corp., USA**
P.O. Box 25252
Santa Ana, CA 92799-5252
949-460-5688
www.kawasaki.com

**Kimball Products**
P.O. Box 792
Benton Harbor, MI 49022
800-358-4586
www.kimballproducts.com

**Kolpin Power Sports**
P.O. Box 107, 205 N. Depot Street
Fox Lake, WI 53933-0107
877-956-5746
www.kolpinpowersports.com
*(three-point hitches, rakes, scrapers, cultivators, forklifts, trailers)*

**Kubota Tractor Corp.**
3401 Del Amo Boulevard
Torrance, CA 90503
310-370-3370
www.kubota.com

**Land Pride**
A Division of Great Plains Mfg., Inc.
P.O. Box 5060, 1525 E. North Street
Salina, KS 67401
785-823-3276
www.landpride.com

**Market Farm Equipment Limited**
RR #1 Dashwood
Ontario, Canada N0M 1N0
519-238-2301
www.marketfarmequipment.com/atv.htm
*(hydraulic and manual dump carts)*

**Mays Trail Equipment & Leasing, LLC**
30 Gun Club Road
Sagle, ID 83860-9345
208-263-4212
877-823-1043
www.atvsnowblowers.com
*(ATV snowblowers)*

**Mill Creek Manufacturing Co.**
34 Zimmerman Rd.
Leola, PA 17540
800-311-1323
www.millcreekspreaders.com
*(manure spreaders)*

**Mountaintop Golf Cars, Inc.**
9647 Highway 105 South
Banner Elk, NC 28604
800-328-1953
www.golfcarcatalog.com
*(golf car accessories, including leaf springs, lift kits, extra seats, cargo boxes and vans, hitches, trailers)*

**Northern Tool & Equipment**
2800 Southcross Drive West
Burnsville, MN 55306
800-221-0516
www.northerntool.com
*(hydraulic and manual dump carts, trailers, seeders, mowers)*

**Palmer Industries**
P.O. Box 5707 US
Endicott, NY 13763
800-847-1304
www.palmerind.com
*(electric personal vehicles)*

**Polaris Powersports**
7763 W. Gulf to Lake Highway
Crystal River, FL 34429
352-795-7996
www.polarispowersports.com

**Pug, Inc.**
Feterl Manufacturing Corporation
411 Center Avenue West, P.O. Box 398
Salem, SD 57058
800-367-8660
www.pugpower.com

**Putnam Hitch Products**
211 Industrial Avenue
Bronson, MI 49028
800-336-4271
www.putnamhitch.com

**Rick's Electric Sports Vehicles**
2517 Mike Padgett Highway
Augusta, GA 30906
866-798-2227
www.rickesv.com

**Roda Manufacturing Inc.**
338 Main St.
Hull, IA 51239
888-214-9914
www.rodamfg.com
*(manure spreaders)*

**Snapper**
535 Macon Street
McDonough, GA 30253
800-935-2967
www.snapperpro.com
*(turf/trail/grounds cruiser utility vehicles)*

**Swisher Inc.**
P.O. Box 67, 1602 Corporate Drive
Warrensburg, MO 64093
800-222-8183
www.swisherinc.com
*(cargo boxes and baskets, carts, wagons, cultivators, mowers, tire chains, rotary power sweepers, pallet forks, loader buckets, plows, lawn vacuums)*

**Timberwolf Manufacturing Corporation**
118 Spruce Street
Rutland, VT 05701
800-340-4386
www.timberwolfcorp.com

**Toro Company**
Consumer Division
8111 Lyndale Avenue South
Bloomington, MN 55420
800-348-2424
www.toro.com

**Workensport**
1310 SE Reed Market Road
Bend, OR 97702
800-816-8002
www.coworkensport.com/usv/FireTrackerjr.html
*(fire protection units)*

**Worksaver**
P.O. Box 100
9 Worksaver Trail
Litchfield, IL 62056-0100
217-324-5973
www.worksaver.com
*(mounting kits, rakes, harrows, discs, box scrapers, seeder/spreaders)*

## TRACTOR AND EQUIPMENT MFRS. AND DEALERS

**AGCO Corporation**
4205 River Green Parkway
Duluth, GA 30096
770-813-9200
www.agcocorp.com

**Association of Equipment Manufacturers**
Hubbard Publishing, Inc.
P.O. Box 525
Windsor, WI 53598
800-369-2310
www.aem.org
*(farm equipment safety manuals)*

**Branson Tractor Company**
2100 Cedartown Highway
Rome, GA 30161
877-734-2022
www.bransontractor.com

**Bush Hog, L.L.C.**
P.O. Box 1039
Selma, AL 36701
334-874-2700
www.bushhog.com
*(mowers, augers, loaders, rakes, blades)*

**Case IH**
700 State Street
Racine, WI 53404
262-636-6011
www.caseih.com

**Deere & Company**
One John Deere Place
Moline, IL 61265
309-765-8000
www.deere.com

**Massey-Ferguson, Inc.**
AGCO Parts Division
North America Headquarters
1500 Raddant Road
Batavia, IL 60510
630-879-3300
www.masseyferguson.com

**New Holland**
500 Diller Avenue
New Holland, PA 17557
888-290-7377
www.newholland.com

**Parma Company**
P.O. Box 190
Parma, ID 83660
208-722-5116
www.parmacompany.com
*(arena groomers)*

**Rhino**
1020 S. Sangamon Avenue
Gibson City, IL 60936
877-408-3297
www.servis-rhino.com

**Ron's Equipment Co., Inc.**
906 N. U.S. Highway 287
Fort Collins, CO 80524
877-495-9482
www.ronsequipment.com
*(new and used farm equipment)*

**ROPS search page** to find a system
to fit your tractor:
www.marshfieldclinic.org/NFMC/rops/default.htm

**U.S. Farmer**
www.usfarmer.com/about/farm-equipment-links.htm
*(extensive list of manufacturers of tractors and equipment)*

## TRAILERS, HITCHES, AND ACCESSORIES

**4-Star Trailers**
10000 NW 10th Street
Oklahoma City, OK 73127
800-848-3095
www.4startrailers.com

**Aluma Ltd.**
101 E. Seneca
P.O. Box 287
Bancroft, IA 50517
515-885-2398
www.alumaklm.com

**American Trailer Mfg. Co.**
8645 Westpark Street
Boise, ID 83704
888-375-8757
www.americantrailermfg.com

**Automotive Accessories Connection Inc.**
5161 Warner Road
Garfield Heights, OH 44125
888-425-2885
www.accessconnect.com
*(hitches, mirrors, brake controllers,
winches, bed rails, battery chargers)*

**B & W Gooseneck Hitches**
1216 Highway 224
P.O. Box 186
Humboldt, KS 66748
800-248-6564
www.turnoverball.com

**Barrett Trailers**
P.O. Box 1500, 1831 Hardcastle Boulevard
Purcell, OK 73080
405-527-5050
www.barrett-trailers.com

**Bee Trailers**
524 Harrell Road
Climax, GA 39834
800-266-2052
www.beetrailers.com

**Big Tex Trailers**
850 I-30 East
Mt. Pleasant, TX 75455
903-575-0300
www.bigtextrailers.com

**Brenderup Real Trailers**
2715 S. CR 1208
Midland, TX 79706
800-745-1306
www.brenderuprealtrailers.com
*(Euro style)*

**Carriage Industries**
P.O. Box 746
Logan, UT 84323
800-742-7047
www.logancoach.com

**Charmac Trailers**
452 South Park Avenue West
P.O. Box 205
Twin Falls, ID 83301
800-544-7904
www.charmactrailers.com

**Circle J Trailers**
Western World Inc.
200 N. Kit Avenue
Caldwell, ID 83605
800-247-2535
www.circlejtrailers.com

**CM Trailer Mfg.**
P.O. Box 680, 200 County Road
Madill, OK 73446
888-269-7577
www.cmtrailers.com

**Cushion Glide Coupler**
Mark Rehme
1449 County Road 1590
Rush Springs, OK 73082-3060
800-324-8032
www.cushionglide.com
*(gooseneck shock-absorbing coupler)*

**Draw-Tite**
47774 Anchor Court West
Plymouth, MI 48170
800-521-0510
www.draw-tite.com

**EquiSpirit Trailer Company**
P.O. Box 1987
Southern Pines, NC 28388
877-575-1771
www.equispirit.com

**Equistar (Merhow)**
306 Depot Street
Bristol, IN 46507
574-848-4445
www.merhow.com/equistar.html

**EZQuip Corp L.L.C.**
P.O. Box 1356
966 Station Branch Road
Prestonsburg, KY 41653
866-394-5824
www.ezhitch.biz
*(EZ HITCH Trailer Guide for one-person hook-up of straight-pull trailers)*

**Featherlite Inc.**
Highway 63 & 9
Cresco, IA 52136
800-800-1230
www.featherliteinc.com

**Gooseneck Trailer Mfg. Co.**
4400 East Highway 21
Bryan, TX 77808
800-688-5490
www.gooseneck.net

**Hart Mfg Inc.**
Highway 81 South
P.O. Box Drawer C
Chickasha, OK 73023
888-810-4278
www.harttrailer.com

**Hitch Solutions**
6405 East Twin Creeks Drive
Idaho Falls, ID 83401
208-221-6281
www.hitchsolutions.com/ranchhitch.html
*(gooseneck and fifth-wheel hitches, hitch adapters, adjustable receiver hitches)*

**Horton Hauler**
101 Industrial Boulevard
Eatonton, GA 31024
800-714-7961
www.hortonhauler.com

**Jamco Trailers**
36 Highway 4
Brucefield
Ontario, Canada NØM IJØ
519-233-7489
www.jamcotrailers.com

**Kiefer Built Inc.**
P.O. Box 88
Kanawha, IA 50447
888-254-3337
www.kieferbuiltinc.com

**Kingston**
182 Wapping Road
Kingston, MA 02364
800-504-3088
www.kingstontrailers.com

**Load Trail, Inc.**
Rt. 2 Box 154-A
Sumner, TX 75486
903-784-8719
www.loadtrail.com

**Max-Air Trailer Sales**
1908 S.E. Frontage Road
Fort Collins, CO 80525
800-456-2961
www.max-airtrailers.com

**Reese Products Inc.**
Cequent Towing Products
47774 Anchor Court West
Plymouth, MI 48170
800-326-1090
www.reeseprod.com

**S&H Trailer Mfg. Co.**
800 Industrial Drive
Madill, OK 73446
800-367-5577
www.sandhtrailers.com

**Scott Murdock Trailer Sales**
6545 ECR 14, Highway 60
Loveland, CO 80537
800-688-8757
www.murdocktrailers.com

**Sundowner**
9805 OK Highway 48 South
Coleman, OK 73432
800-654-3879
www.sundownertrailer.com

**The Used Trailers Company**
1204 B Wagonhammer
Gillette, WY 82716
877-606-0250
www.usedtrailers.com
*(nationwide directory to new and used trailer dealers of all types)*

**Titan**
2306 S. Highway 77
Waterville, KS 66548
866-294-4514
www.titantrailer.com

**Trail-et, Inc.**
P.O. Box 499, 107 Tower Road
Waupaca, WI 54981
800-344-1326
www.trail-et.com

**Trails West Mfg.**
P.O. Box 67, 65 North 800 West
Preston, ID 83263
208-852-2200
www.trailswesttrailers.com

**Turnbow Trailers, Inc.**
P.O. Box 300, 115 West Broadway
Oilton, OK 74052
800-362-5659
www.turnbowtrailers.com
*(reverse slant)*

**Una-Goose**
P.O. Box 787
Versailles, IN 47042
800-992-5099
www.una-goose.com
*(removable ball gooseneck hitch)*

**Universal Trailer Corporation**
11590 Century Boulevard, Suite 103
Cincinnati, OH 45246
800-992-5099
www.universaltrailercorp.com
*(Sooner and Exiss brand trailers)*

**WW Trailers**
P.O. Box 807, Highway 199 West
Madill, OK 73446
405-795-5571
www.wwtrailer.com

## SKID STEERS, WHEEL AND TRACK LOADERS, AND ATTACHMENTS

**Attachment Technologies, Inc.**
P.O. Box 266, 503 Gay Street
Delhi, IA 52223
800-922-2981
www.bradcoattachments.com
*(skid-steer attachments)*

**HST Sales**
22311 Bear Valley Road
Apple Valley, CA 92308
760-240-1087
www.hstsales.com
*(tracked loaders)*

**Mustang Manufacturing Co.**
P.O. Box 547
Owatonna, MN 55060
507-451-7112
www.mustangmfg.com

# Index

Note: Numbers in *italics* indicate illustrations and photographs; numbers in **bold** indicate charts, lists, and tables.

# Other Storey Titles You Will Enjoy

*How to Build Animal Housing,* by Carol Ekarius. Build shelters that meet animals' individual needs with 60 plans for barns, windbreaks, and shade structures plus designs for watering systems, feeders, chutes, stanchions, and more. 272 pages. Paperback. ISBN 1-58017-527-9.

*Renovating Barns, Sheds & Outbuildings,* by Nick Engler. Preserve history and save money by renovating and restoring rather than replacing old barns, sheds, and outbuildings. 256 pages. Paperback. ISBN 1-58017-216-4.

*Building Small Barns, Sheds & Shelters,* by Monte Burch. Extend your working, living, and storage areas by building low-cost barns, sheds, and animal shelters using these easy-to-follow plans and construction methods. 248 pages. Paperback. ISBN 0-88266-245-7.

*Horsekeeping on a Small Acreage,* by Cherry Hill. Thoroughly updated, full-color edition of the best-selling classic details the essentials for designing safe and functional facilities whether on one acre or one hundred. Hill describes the entire process: layout design, barn construction, feed storage, fencing, equipment selection, and much more. 320 pages. Paperback. ISBN 1-58017-535-X.

*Storey's Guide to Raising Horses,* by Heather Smith Thomas. Whether you are an experienced horse handler or are planning to own your first horse, this complete guide to intelligent horsekeeping covers all aspects of keeping a horse fit and healthy in body and spirit. 512 pages. Paperback. ISBN 1-58017-127-3.

*The Horse Behavior Problem Solver,* by Jessica Jahiel. Using a friendly question-and-answer format and drawing on real-life case studies, Jahiel explains how a horse thinks and learns, why it acts the way it does, and how you should respond. 352 pages. Paperback. ISBN 1-58017-524-4.

*Horse Handling & Grooming,* by Cherry Hill. This user-friendly guide offers a wealth of practical advice on mastering dozens of essential handling and grooming skills necessary in maintaining consistent equine well-being. 160 pages. Paperback. ISBN 0-88266-956-7.

*Stable Smarts,* by Heather Smith Thomas. Gathered here in a readily accessible handbook, Thomas's hundreds of useful tidbits — gleaned over a lifetime of working with horses day in and day out — will generally improve and simplify the quality of life on a horse farm, while saving time and money in ways you never thought possible. 320 pages. Paperback. ISBN 1-58017-610-0.

*These and other books from Storey Publishing are available wherever quality books are sold or by calling 1-800-441-5700. Visit us at www.storey.com.*